THE
ENTROPY
EFFECT

AN EXPLORATION INTO SYSTEMS AND ENTROPY ~
THE FINAL FRONTIER OF SCIENCE

Paul E. Triulzi

THE ENTROPY EFFECT
AN EXPLORATION INTO SYSTEMS AND ENTROPY ~
THE FINAL FRONTIER OF SCIENCE

iUniverse books may be ordered through booksellers or by contacting:

iUniverse
1663 Liberty Drive
Bloomington, IN 47403
www.iuniverse.com
1-800-Authors (1-800-288-4677)

ISBN: 978-1-5320-4311-6 (sc)
ISBN: 978-1-5320-4312-3 (e)

Print information available on the last page.

iUniverse rev. date: 04/05/2018

WHEN ONE CONSIDERS THE VARIOUS EFFECTS OF SYSTEMIC FORCES, ONE CANNOT IGNORE THE ROLE OF ENTROPY IN DETERMINING THE RESULT OF A SYSTEM'S FUNCTION. THIS IS THE FINAL FRONTIER OF SCIENCE: TO UNDERSTAND THE COMPLEXITY AND INTERDEPENDENCE OF ENTROPY AND SYSTEMS.

This book is dedicated to my parents, Eugene and Pauline, and to Anne, and Daniel.

CONTENTS

The word ENTROPY is derived from the Greek word "tropē" meaning "change", "transformation" or "evolution".

PREFACE

This is as much a book of questions as it is a book of answers. When you delve into the concept of entropy you realize that the effects are multivariate and very complex; so much so, that to rigorously develop solutions requires a great deal of specialized knowledge, careful examination and advanced mathematical development. This is because entropy as a concept, rather than a strict thermodynamic measure, infiltrates every aspect of every physical system at the microscopic level, and aspects of consideration and observation at the macroscopic level. In many cases, rigorous development cannot be expected from one individual or even from one generation. But to ignore the challenges and benefits of understanding the role of entropy in systems is to ignore, in the author's opinion, seeking the last known "holy grail" of the physical sciences.

At a macroscopic level, we can more easily understand the relationships and trade-offs of entropy in the real world. With this type of analysis, one can arrive at some answers or conjectures, postulate some theories, develop some equations, and pose a lot of questions. It is the author's hope that the questions unanswered in this book will spur researchers, mathematicians, and scientists to close the gap that exists in our knowledge of the interdependence of entropy and systems, and to apply resulting discoveries to make the world a better place.

This book proposes conclusions and hypotheses derived from applying the **concept** of entropy to an analysis of physical and virtual systems. This book is not a treatment of entropy as a thermodynamic measure, but rather a conceptual exploration of entropy's role in systems.

INTRODUCTION

Most of us have at some time run across a person at work who keeps a very messy desk or office, and we think as we see it - how in the world is it possible to get any work done amid such disorder? What we have thought at such a time has a great deal of significance beyond the situational aesthetics. We are acknowledging, perhaps unknowingly, a fundamental principle that will be amplified and iterated in this book that as the disorder of a system increases, the amount of work capable by that system per unit of energy generally decreases. This may be common sense to most, but it is also a key principle that helps to govern nature, the behavior of humans and society, and a variety of phenomena whose behavior is not obvious and taken for granted.

Some have on occasion commented on the messy office by saying something like: "How can you work in this office?" or "How can you find anything on your desk?" To which, a clever reply often parallels: "A clean desk is a sign of a weak mind." Of course, this does not and should not imply that a clean desk means you have a weak mind. But what it does mean is that to function effectively with a messy desk, you had better have a strong mind, or more precisely, a very good memory and a strong mental sense of order. This is because the mind can virtualize the order of the messy desk. The messy desk is still in a state of physical disorder, but the mind has created a virtually ordered system with less disorder than the physical messy desk.

It is therefore reasonable to assert that the "virtual desk" system ("strong" mind applied to messy desk) has less overall entropy than the "physical desk" system (messy desk by itself) even though part of the virtual desk system is not physical but exists as thought or as knowledge. The ability of humans to reduce entropy through thought and the propagation of knowledge is a key tenant of this book, and an ability that may be mostly unique to humans as a species in the animal kingdom.

Two persons can produce equal results dealing with unequal amounts of entropy, one discriminating factor being the algorithms used to produce the results and the energy or time expended. In many cases, less entropy is better, but some algorithms may not function as well with less entropy, because some algorithms are developed to take advantage of a particular level of entropy.

This book examines what happens to systems when the entropy of the systems varies as the work is being done. This book will lead to some conclusions that should be taken as hypotheses meant to challenge scientists, researchers, and philosophers to develop new techniques for solving problems, and new theories to explain the physical world and the world of thought.

Some engineers and scientists believe that the concept of entropy should be applied only in its original thermodynamic role of heat transfer and work. Others are comfortable with expanding the concept to a wider non-thermodynamic understanding. I agree with a wider application of the concept of entropy and this book attempts to broaden the application horizon. Furthermore while entropy is a quantity that was initially developed to explain heat energy loss in thermodynamic processes, this book accepts the more general approach of a statistical representation of entropy put forth by Boltzmann, and of the more colloquial understanding that the term "entropy" can be used to describe disorder in real or imaginary systems.

The concept of entropy is embedded in almost everything we live and work with, as well as within ourselves. It is hard to understand why it has not been considered more of a main component of systems analysis and emphasized more equally with other physical measures such as mass, speed, force, acceleration, and time. Perhaps because entropy is so much a part of the fabric of the physical world, it escapes analysis much as does the canvas of a famous painting.

My hope is that this book will pose questions and pave the way to an illumination and interest in entropy, and how understanding it can shape the success and health of our lives, produce new discoveries, and help us wisely use the earth's resources.

CHAPTER I

WHAT IS ENTROPY?

There are numerous variations on definitions for the concept of Entropy based on the fields in which a definition is applied. However, there is agreement on three fundamental derivations for this concept, which are explained below.

Definitions of Entropy

There are three scientifically accepted derivations for entropy in current practice. The original is based on the laws of thermodynamics, the second is based on a statistical consideration of molecules in ideal gases, and the third is derived from considering the information capacity of a stream of binary digits.

First Derivation

The original derivation of the concept of entropy comes from observations of idealized reversible heat engines and was put forth in 1850 by Rudolf Clausius. This defines entropy as the quantity of a system's thermal energy unavailable for conversion into mechanical work. "As a physical system becomes more disordered, and its energy becomes more evenly distributed, that energy becomes less able to do work."[1]

Clausius derived his concept of entropy from the study of the Carnot heat cycle. In a Carnot cycle, heat Q_1 is transferred from a 'hot' reservoir at temperature T_1, to a colder reservoir with heat of Q_2 at a lower temperature, T_2. Clausius saw that there is an inherent loss of usable heat when work is done, and he termed this loss *Entropy (S)*. This observation was designated the Second Law of Thermodynamics which states that a change in the entropy *(S)* of a system is the transfer of heat *(Q)* in a closed system driving a reversible process, divided by the equilibrium temperature *(T)* of the system. [2]

Specifically, this definition is expressed by

$$dS = \delta Q/T$$

As heat is transferred from one reservoir to another to do work, the Second Law says that the total entropy of the two reservoirs must increase, otherwise no work can be done.

Second Derivation

The second derivation of the concept of entropy comes from observations of molecular and atomic systems put forth by Boltzmann. This definition treats entropy as a broader statistical phenomenon that is measured by the amount of disorder or randomness in a closed system. Specifically, this definition is expressed by

$$S = k \, Log \, P$$

where *P* is the probability that a particular state of a system exists, and *k* is the Boltzmann constant. What Boltzmann showed is that if we measured all the possible states that a system could have, then the entropy of a system in a particular state would be proportional to the probability of that particular state occurring. While traditional thermodynamics does not embrace this definition, it has been shown that the thermodynamic definition is derivable from this statistical definition. Therefore, the Boltzmann derivation is considered today a broader and more encompassing definition and a sound mathematical representation of the concept of entropy.

Third Derivation

In the 20th century, a computer scientist at Bell Labs, Claude Shannon, added to the treatment and definition of the concept of entropy by applying this concept to the transmission and usefulness of information. Shannon theorized that the more random a stream of data, the greater is its potential for carrying information. The more predictable a stream of data is, the less information is available in that stream. Though somewhat counterintuitive, the reasoning is that the more predictable a stream of data, then the more redundant is its information, reducing information capacity and therefore making it less useful.

If a message sent is entirely predictable to the receiver, then it contains no useful information, because the receiver already knows the content of the message before it has been sent. Shannon quantified his theory by measuring the randomness of a stream of data in terms of the probabilities for each bit of data. His equation for disorder based information potential is:

$$S = -\sum p_i \log_2 p_i$$

where p_i represents the probability of a particular data message among all the possible messages the data could generate. If simplified, Shannon's information entropy equation becomes the same as Boltzmann's statistical entropy. The similarity is not coincidental, and so our current understanding of entropy now includes more than just the measurable thermodynamic evidence or the statistics of molecular behavior.

Arrow of Time Concept

By the Clausius definition of entropy, every time we use energy to do work some of it is being lost to entropy. By using energy, we are apparently increasing the overall entropy of the universe irreversibly. It is commonly accepted by scientists today that the entropy of our universe is increasing every second of every day. Terms such as "heat death" are used to describe the end result of this apparent progression where molecules reach maximum uniform dispersal and as a result contain no useful energy.

In fact, we humans don't need to expend energy to increase disorder or molecular dispersal. Nature is happy to be the agent for this as systems left to themselves experience increased disorder. For example, a parked car that is not used will begin to rust, the engine will begin to seize, the gasoline will evaporate and varnish the fuel system, the air in the tires will escape, and it will become unusable as a transportation system. The car is changing from an ordered structured design towards a state of disorder that eventually renders it non-functional. It is not hard to understand from the simple example above that systems left to themselves tend toward increases in disorder when energy is not applied to maintain or increase order.

According to Boltzmann's analysis, the tendency of systems to randomly decrease in disorder is for all practical purposes statistically impossible. His analyses show that more ordered physical states based on random action are so statistically remote that they are mostly impossible.

Since the use of energy to do work increases entropy per the second law of thermodynamics, and since inanimate systems left alone become more disordered over time essentially because their structures unravel, the concept of entropy as a measure of time arises. As time passes, universal entropy increases and specific system entropies also increase. From this notion, the passage of time can then be understood and even measured in terms of units of entropy increase. The concept of the "Arrow of Time" always pointing in a forward direction comes from the understanding that entropy increases in the same direction as time. For example, a concentrated grouping of gas molecules placed in one part of a closed container will result over time in the molecules dispersing uniformly to every part of the container. As time passes, the molecules become more dispersed and disordered and although theoretically possible, are not statistically likely to reassemble themselves into the same small group in one part of the container. As time passes, the system tends towards maximum entropy. When the system has reached its maximum entropy, we could say using an entropy clock that time has stopped in the system; the entropy clock will not reverse without the intelligent application of energy we call work.

How This Book Defines and Considers Entropy

The Clausius, Boltzmann and Shannon formulas involve the thermodynamic and mathematical treatments of entropy. Such analysis of entropy is not my intent in this book. Rather, I am more focused on the concept of entropy. Consequently, this book uses a general and conceptual definition for entropy. This is not to say that mathematical proofs are not extensible from the concepts argued, but that others may tackle those derivations.

For the purposes in this book, the term entropy is used in accordance with the three descriptions but mainly as a measure of disorder, or the lack of order, in a closed system. A closed system is defined as a bounded physical construct. A physical construct is defined here to be a 3-dimensional form existing in reality. Disorder is defined as the amount of randomness in a system or as the amount of absence of order or structure among the components that make up the system. Order is defined as a describable and deterministic arrangement of the components of a system.

These definitions go to my intent in this book, which is to generate some new observations and theories involving entropy as the main factor of consideration. While the discussion varies in the degree of formalization and could be further formalized in mathematical terms, there is little attempt to do that here. Some formulaic proposals are put forth but not proven. My main interest is to offer a discussion of ideas around the concept of entropy and systems. Some of the conclusions around this discussion will hopefully give rise to further consideration and development of mathematical treatments.

~ ~ ~

CHAPTER II

ENTROPY, ENERGY AND WORK

There is a fundamental relationship between the entropy of a system and the amount of work a system can do with a given expenditure of energy. This fundamental relationship is a key to understanding fully both the simplicity and complexity of systems. Let us begin.

Entropy and Work in a Closed System

If we enter a teenager's bedroom and close the door, we could assume that we are in a closed system if we define the walls and door, floor and ceiling of the room as the boundaries of a physical construct.

Furthermore, if the room is neat and orderly, we can say the entropy of this physical construct is low. If the room is untidy but orderly, we can say that the entropy is average; likewise if it is tidy but disorderly. If the room is untidy and disorderly, we can say that the entropy of the room is high.[3]

Why is this important? Why do some people make such a fuss over a messy room? If you asked them, they might say something like: "We want the room to look nice" which reflects their aesthetic concerns; or "How can you do your homework with such a mess", which reflects their functional concerns; or "How can anyone stand to live in a room like this", which reflects their hygiene or psycho-social concerns.

The last statement is the most revealing statement about the psychology of an individual. Some people like to have a messy (disorderly) room while others cannot stand the messiness. Some don't care. Each person has their own preference for the space they live in and how it should be structured. For some teenagers, a messy room may not be a problem because it suits their functionality needs just fine. Their room is a personal space, and the messiness (disorder) may give the room a friendliness that is uniquely their own, or that order is not important at this stage in their life. You could argue that the messiness (or disorder) gives the room an artistic quality and that it makes one feel creative.[4] Other teenagers may not want a messy room, but they also don't want to do the work necessary to keep their bedroom in order. It takes energy and time and work to keep a bedroom in order. In summation, the order of a room says a lot about the psychology of the person living there that may not be relevant to the subject of this book, but, it is a good example to use to begin the discussion of entropy, energy and work. Disorder may give a room the feeling of warmth, creativity, or uniqueness but what does disorder take away from the room when one thinks of it as a functional system?

Say you are in your bedroom studying without a computer and your mother comes in and says: "Could you look up the spelling of a word for me?" In a tidy ordered room, you would reach over to the bookshelf and run your finger across the row of books to the Dictionary, pull it out, look up the word alphabetically, and give the spelling to your mother. The effort expended would not be excessive and, just as importantly, the time required to give your mother the answer would be reasonable enough that she would consider it productive use of her time and yours to ask you for the spelling of the word.

On the other hand, if it took you 5 or 10 minutes to find the correct spelling for the word, your mother would most likely not come to you next time for a similar thing because she would feel the time spent on her request was unproductive for her and you.

To fulfill the mother's spelling request in a messy disordered room, you would start moving stuff around to uncover a stack of books, look through the stack for the dictionary but not find it, then look through a different stack of books, etc., until you find the Dictionary. It might take you a short time to put your hands on the dictionary (low probability), or it could take you a long time (high probability), which essentially means that the functioning process involves a higher degree of uncertainty. It is also possible that you may not find the dictionary in a reasonable enough length of time and the spelling request would be annulled. If you found the dictionary, you would look up the word alphabetically (as in the previous example), and give the spelling to your mother.

In general, the time and effort to get the correct word spelling for the mother would be probabilistically greater in the messy room than in the neat and ordered room. Also the amount of work performed and energy expended to get the correct spelling would be greater in the messy room than in the ordered room. All the items in both tidy and messy rooms are the same, but in each room scenario the work performed is different.

A third scenario to consider is a messy room but with a very mentally ordered occupant such that locating the dictionary is as easy and quick as if it were neatly placed in order on a bookshelf. The dictionary may be under a stack of papers, but the occupant knows exactly where it is and can produce it immediately. In this case, the work performed and the time spent differs only slightly than in the ordered room. While the room is not physically ordered, the occupant has mentally created a virtual order, but this order is not evident and is not physical.

We can also make another observation from the room example. Getting the correct spelling involves three entropies, one being the state of the room's contents, the second being the state of the occupant's mind, and the third being the state of the dictionary. Each of these three things are subsystems to the room system and have their own state of order that together affect the work output of the room as a bounded larger system. From the mother's perspective, the bounded larger system is the room she stands outside before she walks through the door. She needs the answer to a spelling question and she knows the answer is available in that room. She doesn't necessarily know or think about all the subsystems that come into play to produce her answer. But from the occupant's perspective, providing the answer to the spelling question is dependent on the order of the room, the occupant's state of mind, and the order of the dictionary.

The mother observes the work output of the whole system in getting her answer. Does the room produce the answer to her spelling question in a timely and efficient manner, or should she go elsewhere to get her answer. The room occupant perceives the work output as depending on the state of three subsystems: 1) occupant's state of mind, 2) the order of the room, and 3) the order of the dictionary.

From the mother's perspective, the closed and bounded system of the room that gives her a quick answer to her spelling question has an entropy which we will call E_r ("r" for "room"). From the occupant's perspective, answering the spelling question depends on the entropy of specific subsystems within their room, mainly for this analysis, the arrangement of the contents of your room (E_c), the entropy of the occupant's mind (E_m), and the entropy of the dictionary (E_d).

The order of the room's contents refers to how easy is it to locate a specific item in the room, in this case the dictionary. The occupant's state of mind refers to the mental memory, mental focus, and psychological willingness to answer the spelling question. The order or structure of the dictionary affects how easy it is to look up a desired word. Answering the spelling question requires the mother depend on three subsystems that affect the ability of the "room system" to produce an answer quickly and efficiently.

We could summarize the above example by stating the system's entropy in terms of its subsystems: $E_r = E_c + E_m + E_d$ where the "+" operator does not imply simple addition. This example illustrates the difference between macroscopic and microscopic entropy, the room as a whole macroscopic system with entropy E_r, being the result of "microscopic" sub-system entropies. It is true the room contains other subsystems that may or may not contribute to E_r, but only the main ones that matter towards answering the spelling question have been considered. If your mother always viewed your room as the place to get answers to spelling questions, then E_r would tell her how well your room can deliver the answer. The higher the value of Er, the less probable it is that she will get the spelling answer satisfactorily in a timely fashion[1].

If there is a sibling in a separate room where the mother could also go for the answer to her spelling question, she may find this other room to be more productive to getting her answer. The room with the lowest system entropy (E_r) would most likely be the best choice for getting the answer to the spelling question, assuming both rooms are equally accessible to her. While the macroscopic entropy helps us to make a choice, it is the subsystems within E_r that make the difference. The work output of large systems is based on the entropies of their dependent subsystems, all other factors[5] being equal. This both simplifies and complicates the analysis of systems and their outputs.

Entropy and Work in a Subsystem

To explore the idea of subsystem entropy, let's look specifically at the dictionary subsystem used in the room example above. The dictionary is a key component to answering the spelling question and completing the work that needs to be performed by the larger room system. While the dictionary on its own does not complete the task, getting the task done is dependent on it. How does the entropy of this subsystem affect the energy consumed and the timely completion of the task?

A dictionary is a set of words ordered alphabetically to make it easier for us to find the word we are looking for[2]. The greater the order of the

---

[1] "in a timely fashion" refers to information being cogent, i.e., made available at the right time for it to be useful. There is a window for this, or a "time set" in which the information is useful. Also, the speed at which the information is presented carries its own parameter or effect. A quick answer is more desirable, and carries more impact than a slow answer. We can think of information in this way as having momentum, which equates to the <u>value of the information</u> times the <u>speed at which it is delivered</u>. See the chapter on Information Entropy for more treatment of this concept.

[2] It is worthwhile to point out here that the efficacy of alphabetical ordering is dependent on the observer's knowledge that the alphabet in English goes from A to

dictionary, the lower is its entropy. In fact, the dictionary subsystem has a characteristic entropy or structure that defines it. Its structure defines the type of the dictionary and the algorithm to use for looking up a word. The dictionary's entropy is a result of its algorithmic[3] entropy and its structure[4]. The dictionary's entropy is built-in by its construction and layout, where pages of words are assembled in a particular way, and they remain that way after the book has been published. The entropy of the dictionary subsystem is definable because of its human development and it is relatively static with time.

To further explore the order of a dictionary, let's first consider what it would be like if it did not have an alphabetically ordered list of words? The dictionary would then have the same physical structure as a normal dictionary (bound and numbered pages, etc.) but the order of words would be in non-alphabetical order, resulting in a high amount of entropy for an alphabetical algorithm. So if one opened the dictionary to the first page, the words would not be ordered alphabetically and would not necessarily begin with the letter "a" as expected. If all the pages had words in random order, how much effort and time would one expend to find a specific word? One might get lucky and find the desired word in the first few pages. But most likely, one would have to look through page after page for the desired word, consuming an inordinate amount of time and energy. For anyone seeking the spelling of a word, this type of dictionary would be very hard to use, even though it contained all the desired information. The disorder or entropy of the dictionary contents is so high that accomplishing the task of finding a word is probabilistically so difficult that the subsystem has little functional value.

In a random dictionary (no pun intended to), the word sought could be anywhere in the dictionary's content of about 50,000 words and could take

---

Z in a particular order. But a different arrangement might be more effective.

[3] Algorithmic entropy refers to the "logic" or "intelligence" built into the arrangement of words in the dictionary. Dictionaries can be ordered alphabetically, or in some other order that might be more useful. The type of order employed helps determine the system's entropy.

[4] Structure determines static entropy. Highly structured frameworks have lower entropy.

a very long time to find. The task of finding a specific word in a randomly ordered list of 50,000 words might be considered too time consuming to be reasonable or worthwhile. Without more order (lower entropy) in the dictionary, the work and time[5] to find a word is certainly unreasonable, though not impossible. The mother experiencing the room system waiting for the spelling of a word would most likely give up and not use the system again, even though all the subsystems are inn place. To recap this situation, the disorder of the room contents and of the occupant's mental state may be very low, but the desired task cannot be reasonably accomplished because the entropy of the dictionary subsystem is very high.

To improve the situation we want to lower the entropy of the random dictionary. If the words in the dictionary were partially ordered, say in four groups of five starting letters and one group of six[6], it would immediately take less time and work to find the word you want than if there were just a random order, but it still would not be as easy as using a fully ordered dictionary. In our partially ordered dictionary, the time and work might still be too much to find the correct spelling you are looking up[7], but it would be less than using the random dictionary. Additional order (lower entropy) improves using the dictionary. If order is increased such that words in the dictionary are grouped by first letter, but not by a second letter, or a third etc., then all the words starting with a particular letter would be together in the dictionary in one group. Finding your word would still take some time, but this dictionary with 26 groups of words would be more efficient than the dictionary with words ordered into just

---

[5] Probability would allow you to calculate expected search times based on the number of words in the dictionary and reading speed.

[6] In this type of dictionary, words in group 1 would have starting letters of a,b,c,d,e; group 2 would have starting letters of f,g,h,i,j; etc. Within each group, the words would still be in random order.

[7] It is important to note that there is a probability component to finding the word you seek in a random large collection, or a partially ordered collection of words. It is probable, but highly unlikely, that the word you seek is, for example, on the first page of the dictionary. In this case, the time and effort to find the word would appear to be minimal. But it is more probable, that the particular word you are looking for is buried on some other page in the dictionary, and the time and effort to find the word is substantial.

five groups. So as more order is applied to the dictionary, less energy and time is required to complete the task of finding a particular word. If the dictionary were further ordered, it becomes clear that as the entropy of the dictionary decreases, the time and effort to use the dictionary successfully to find a word also decrease.

We can see that many different types of dictionaries are possible, each with a different type of order or disorder (entropy). But also, and very importantly, what defines what we call a "Dictionary" is its particular type of order and the associated entropy and algorithm for using it. If the entropy of a dictionary rises above a specified threshold, then the dictionary no longer looks or functions like "the Dictionary" that we expect, and the algorithm for using the dictionary may become ineffective beyond a certain entropy level. The very nature or characteristics of the Dictionary, therefore, are dependent on its entropy. This simple example can be applied to all systems, inanimate or animate. This has profound implications for the analysis of system characteristics and function.

Since there can be many forms of order in a dictionary, each with its own entropy and algorithms, it is reasonable then to state that a Dictionary's nature is not dependent on a singular entropy, but on a range of entropies measurable for each form or type, as long as the dictionary function as specified by its algorithm still produces a desired output with a reasonable expenditure of energy. Within this range or "band" of entropies for the Dictionary, we can introduce the concept of a nominal entropy, or Entropy Equilibrium (EE). The EE is the entropy at which the dictionary is closest to the nature of the Dictionary system as we know it. At the EE, the Dictionary is in its "natural" or "characteristic" state, that is, the state intended for this system. Other dictionaries with different entropies may be available and useful, but the dictionary that we are accustomed to has a designated entropy of EE. Other useful dictionaries have entropies that fall within the band of entropies around the EE. The EE is the point at which a system is most like what is intended[8] and the Entropy Band

---

[8] This opens up an entire area of scientific and philosophical consideration around what is "expected", what is "intended", and what is "provided". Different mindsets (ie, cultures) may select different design algorithms for the function of a system and

(EB) is the range of possible entropies for a particular system to retain its algorithmic function.

When the entropy of our dictionary system has either increased or decreased sufficiently to be outside its EB, then it is no longer a "Dictionary", and its function has been abrogated. In other words, the work output of the dictionary system is no longer commensurate with the input and a given expenditure of energy. We can summarize this further by saying that if the output of a system is no longer valid for a particular input and expenditure of energy, then the entropy of the system has shifted far enough to be outside its EB, and the system no longer possesses its original or "true" nature. The shift in entropy can be due to changes in structure as well as accumulation of non-essential physical factors. A random dictionary that is merely a collection of words bound in the context of a book has physical structure but it cannot function successfully as a dictionary if the words are randomly ordered. It is even less likely to function if the pages are unidentifiable instead of being bound in one order.

Returning to the thought experiment of the dictionary as a subsystem within a room, if the dictionary is outside the EB for dictionaries, and as a critical subsystem within the larger room system, it renders the larger room system inoperable from the mother's point of view, which is that the system will give her an answer to her question.

This simple analysis can be applied to all systems, natural and man-made, and therein is the challenge and the reward of understanding and managing entropy successfully.

---

thus create a different EE. Nature provides us with complete systems but man also designs and develops many systems of his own. For example, in the food industry, we design systems with "artificial ingredients"; and in the agricultural crop industry, we design systems of "genetically modified foods" or "hybrid" plants, which will have a different EE from the "natural" plant.

Principles of Work and Entropy in Systems

If one examines a physical system and measures the amount of work that the system is able to produce with a given amount of energy, the work produced by the system is generally inversely proportional to the amount of entropy present in the system. Specifically, this inverse relationship exists only within a bounded continuum of entropy or the EB. If the entropy lies outside the boundaries of the continuum, then the system cannot produce the expected work, it may be non-functional, or its function may be different. Simply adding more energy may not restore the system to its proper EB, but instead will contribute more to heat and increases in entropy of the super-system[9].

If energy is applied to a system that does work, then the amount of work produced will vary inversely with the amount of entropy in the system. For a given energy input, the entropy-work curve will be continuous up to the thresholds of the entropy band. At the thresholds, the entropy-work curve becomes disjointed. Beyond the threshold points, the entropy-work curve approaches zero work more rapidly than the curve exhibits within the EB. In other words, as entropy increases, work output decreases in a continuous fashion up to the EB thresholds, after which, the system changes its nature and the deterioration of work output ceases or the decrease is much more pronounced for each increment of system entropy change.

Let's examine the above claims by using for our closed physical system a non-inflated balloon made of rubber attached to a nozzle that can be opened or closed. The rubber surface of the balloon and the nozzle form the boundary of our closed physical system. When there is no air in the balloon, the physical system is not capable of performing any work, which in this case, is the capability for the balloon to fly around a room via the expulsion of air through the nozzle.

---

[9] We must theorize that every system is surrounded by a larger system. The entropy of the larger (or super) system is dependent on the entropies of its subsystems and its own resting state of order. The resting entropy is the entropy the system has when zero energy is flowing in or out of the system.

In a balloon with no air, entropy is very low and if we open the nozzle, the system just stays where it is. The entropy of this balloon system with no air inside is very low because there is no entropy associated with air molecules in the system, and the structure of the balloon is relatively static. The entropy of the balloon in this state lies outside the boundaries of the entropy band that we can expect to produce work. The entropy of the system is outside the range where energy, say in the form of heat, if applied to the balloon system, will produce work from the balloon should we open the nozzle.

If there is no air in the balloon, increasing the heat energy on the balloon does not produce the desired work but would just melt the rubber skin, destroying the physical structure of the system and leaving it in a state where the nature of the balloon (an elastic sphere capable of holding air under pressure) has changed cannot be reconstituted. The entropy of the balloon is too low for it to do any useful work.

Consider the original empty balloon system and let us increase the entropy of the system by filling the balloon with some air. Adding air to the balloon is increasing the entropy of the system, because air molecules are now part of the system, and as a gas, they are moving in a random motion inside the closed system. The subsystem consisting of the molecules of air have a relatively high amount of entropy. The entropy of the physical structure has also increased since the balloon's surface area has expanded. The overall entropy of the balloon system has increased to be within the entropy band that is characteristic of a balloon, and we anticipate that it will do work. The balloon with air inside is exhibiting its true nature or intended characteristic that is desirable for it to produce work.

If we open the nozzle of this balloon system and let the air rush out of the balloon, it would propel itself by the force of the rushing air and thus produce the desired work. Once the balloon stopped moving, its entropy would be outside the EB. However, if we closed the nozzle and gently warmed the remaining air inside the balloon, the entropy of the air and balloon structure would increase most likely sufficiently to allow it to perform work again. The applied heat energy expands the residual air by exciting the air molecules sufficiently to press harder against the rubber skin or boundary of the balloon system. If we then open the nozzle, the balloon would again exhibit work. Whether it exhibits the same amount of work depends on the pressure of air inside the balloon. What we can confidently say is that this process of extracting work from the balloon cannot be repeated again and again[10], because eventually the amount of air left in the balloon will be so small that the heat energy required to increase the air pressure sufficiently will instead melt the rubber skin of the balloon and thus lower the structural entropy below the EB for this object to work.

Let's look briefly at the opposite case where we put too much air in the balloon. As more air is added, the number of gas molecules increases, and the system' entropy increases. Eventually, too much air will rupture the balloon skin and the structure will become so disorganized it can no longer perform any work. Heat energy added to the system will not result in more work. The system is outside its EB. So we see that a system producing work depends at least on structure, algorithm, energy, entropy and time for a system's relevant subsystems.

Taking the balloon example further, let's imagine different states for the rubber skin and air subsystems, and conjecture the following conditions for a balloon that contains an amount of carbon dioxide gas: 1) the system is filled with CO_2 at a temperature that allows the rubber to expand significantly; 2) the system is allowed to rest at room temperature; 3) the system is cooled to 30F degrees, 4) the system is at room temperature with no CO_2 inside, 5) the system is warmed to a point where the CO_2 gas expands the balloon until the rubber skin ruptures. Of these five states, only in states 1) and 2) can the balloon produce the intended work. We can say that the entropies of the balloon in states 3), 4), and 5) fall outside the EB of the balloon system and do not produce the expected work. State 5 is one where either too much or too hot CO_2 gas exists in the system for the system structure of the balloon to remain intact and contain the

---

[10] Each iteration would require slowly applying heat over a longer period of time, a factor worthy of further examination and conjecture. Time is a friend of entropy in that with sufficient time, entropy can be modified to suit the requirements of a system. With very short time frames, entropy change may be limited.

pressure of the gas.[11] In State 5, the entropy of the CO_2 has risen to a point high enough that the system's total entropy is above the upper limits of the EB. Similarly, in states 3) and 4), the cooled CO_2 (or lack of CO_2) inside the balloon has such a low entropy that the entropy of the total system is below the lower limits of the EB. Only when the entropy of the system is within the EB can the system produce intended work, i.e., function in the way and for the purpose that it was designed.

To describe more fully State 1, if we were to not let the gas out but apply a gentle amount of heat energy to the gas in the closed balloon system, the heat energy would warm the air causing the molecules to increase their random motion, which increases the entropy of the gas and the air pressure in the balloon. The entropy change of this closed system is a net decrease because while the molecules are sufficiently more random, they are balanced by the decreasing structural entropy. The work capacity has also increased through the absorption and retention of heat energy in the gas. The entropy of the balloon system has changed but it is still within the band that allows for work to be produced. If the nozzle is opened, the rush of heated gas sends the balloon flying more energetically than the balloon of State 2, producing more work even though both systems in states 1&2 have the same matter and mass.

Consider the scenario for State 5, where the balloon with the CO_2 is

---

[11] Here we have a consideration of the structure of the balloon, ie the rubber skin, having a certain entropy, and the contents of the balloon, ie the gas, having another entropy, and the system's entropy being a combination of the two. The balloon structure is low entropy and the gas is high entropy. Low entropy in structure controls the potential energy of the high entropy gas. If enough gas is forced into the balloon (or made hot enough) to rupture the structure of the balloon, then most of the energy available for work is lost to the environment. By raising the entropy of the gas too high, we are forcing the system to leave the entropy band within which the system produces the desired work. Likewise if we chill the gas too much or remove too many gas molecules, we reduce the entropy too much for work to occur. A stretched balloon has less entropy in the rubber structure than a relaxed balloon, but the entropy of the system as a whole increases as the gas pressure increases. The entropy of the gas subsystem is increasing faster than the entropy of the rubber structure subsystem is decreasing.

slowly heated to the point that the gas pressure bursts the balloon skin. As the gas is heated, the entropy of the system goes on increasing. As long as the skin is intact, the balloon is still able to do work when the nozzle is opened. But if the nozzle is not opened, the increase in entropy caused by the heated gas will eventually force the balloon to burst, and the skin will lose the physical structure that defines it as a balloon. At the moment of bursting, the total entropy of the balloon system has exceeded the entropy band that defines the system and allows its intended work to be produced.

In the above balloon scenarios, only when the entropy of the closed system is within the entropy band that defines the system is the system able to produce work, i.e., converting the potential energy of the air mass constrained by the structure of the balloon skin into work. If the total system entropy gets too low or too high, the expected work will not be produced. If the system entropy stays within the EB, the addition of discrete amounts of heat energy should produce more work.

This is one simple example to illustrate what occurs in every physical system in creation and why entropy is the key to understanding how and when systems work as designed. We assume that these basic concepts apply to physical systems designed and built by man, to biological systems that are present in nature, to larger physical systems in the universe, and to atomic structures that are the building blocks for matter, energy, and force.

Increasing physical entropy can generate energy beyond the energy used to effect the increase through the breakup of structure.

Work and Dependent Subsystems

In the Introduction to this book, mention was made characterizing system disorder as a messy desk, and that it would take a strong mind to get anything done with a messy desk. This of course does not mean that clean desks imply weak minds, only that messy desks require strong mental (aka, virtual) organization to provide the same work output for a given amount of energy that an organized desk would provide. Neat desks can be attributed to strong minds just as easily as they can be attributed to weak minds (low mental organization), but it is clear that physical order (neatness) demands less virtual order.

If a system is in disorder (the messy desk), it takes an ordered system (strong mind) acting in concert to make the whole system capable of work. In other words, we once again propose that the sum of the entropies of two dependent systems (organized mind and disorganized desk) has a resultant entropy that dictates the effectiveness of the combined system. The order of one subsystem can make a related disordered subsystem functional, even though both subsystems are distinct and have their own individual entropies. The ordered mind can apply virtual order to the disorganized desk, and the organized desk can apply order to the disordered mind. These are commutative at a point in time. However, when work is accomplished over a span of time, then the commutativeness of the relationship may fail.

Most systems have dependent subsystems such that the work produced by the system requires the interaction of dependent subsystems. In the balloon example, there are at least three subsystems - the rubber skin, the trapped gas, and the nozzle. The work output of the system depends on the entropy of each subsystem and the efficiency of their interaction.

In this regard, let us consider a branch of the armed forces where a distinct chain of dependence exists between two fairly autonomous groups that are each commanded individually. For each autonomous group we consider a platoon which has its own commander. The platoon could be thought of as a subsystem of a brigade with a certain amount of order and an entropy value that can be measured. Another platoon with a different commander has a significantly different amount of entropy. If the entropies of the two platoons are significantly disparate, how would the platoons affect each other if they were together in combat and had to depend on each other to achieve a combat objective for the brigade?

Suppose platoon O is much more ordered than disordered platoon D, and that their combat objective is to take control of a hilltop fort; but to do that, they must first secure a ridge. Further assume that one of the platoons will be tasked with securing the ridge. If platoon O must wait for platoon

D to secure the ridge before advancing to the fort, then the efficiency of platoon D will affect when (and likely how) platoon O advances to take the fort. The outcome of the mission is judged by the efficiency of taking the fort. The mission efficiency is a combination of the efficiencies of each platoon. The platoons are subsystems that operate independently but are dependent to each other in the brigade system. If platoon O is responsible for seizing the fort beyond the ridge, it cannot do so until platoon D has secured the ridge.

Commutativeness of Dependent Subsystems

In the previous example, we assume platoons O and D are identical in their resources and numbers. What differs is that one platoon functions with greater internal order than the other. As a result, the disorder of platoon D can make securing the ridge consume more time and energy than platoon O would use for the same objective. Platoon D and thus makes platoon O appear less effective in gaining the fort. The outcome of platoons D and O will be less effective than if there were two platoon Os and more effective than if there were two platoon Ds.

When looking at the dependency of subsystems, we can simplify the analysis by putting a constraint on time. For the brigade example, the time to take the fort, which is the objective work of the system, can be specified as a constant. Whichever platoon goes first, the work must be accomplished within a fixed period of time.

If the roles of the platoons are reversed and platoon O takes the ridge much more efficiently and quickly than platoon D did, then platoon D is helped in their advance on the fort by providing more time to accomplish their task, hence platoon D appears more effective than it actually is in securing the fort.

Is the dependency of these two subsystems in achieving the objective strictly commutative? If platoon O is dependent on platoon D for their combined objective, then their outcome is less certain than if platoon D is dependent on platoon O. Why? Platoon O going first gives platoon D more time and therefore a greater potential chance for success. But because time is required for success, Platoon D going first gives platoon O less time and a greater potential chance for failure. The implication is that combining the entropies of two subsystems may not be a commutative operation where time is fixed. For subsystems working concurrently, we may assume that the combination of entropies is commutative, but for subsystems working consecutively in a timeframe, the combination of entropies may not be commutative for the overall system.

One might assume that the ideal situation is to balance the entropies of all the subsystems to achieve optimum productivity in the larger system. Will an overage or underage of entropy in one or more subsystems additively affect the total entropy, or will the resultant system entropy be unequal to the sum of the component entropies?

We should also consider whether a very high entropy subsystem matched to a very low entropy subsystem produces system entropy that matches the combined entropies divided evenly between the two subsystems.

Assume you have two sets of dependent subsystems. Algebraically, if $e1$ is the entropy for subsystem 1 and $e2$ is the entropy for subsystem 2 and $e1$ and $e2$ are dependent[12], and similarly for $e3$ and $e4$, then can we say that if $e1 > e2$ and $e3=e4=(e1+e2)/2$, then $w(e1,e2) < w(e3,e4)$, where w is work output as a function of the entropies of two dependent systems? In other words, is it valid to state that the greater the delta between the entropies of dependent subsystems, the lower is the efficiency of the total system compared to the efficiency of systems where the entropies are more similar. One can also consider the following for subsystem entropies:

if $|e1 - e2| \gg |e3 - e4|$, and $e1+e2 = e3+e4$, then is $w(e1,e2) \ll w(e3,e4)$?

Is it true that two subsystems needing to function together to produce work are more efficient if their entropies are more similar? If so, we could

---

[12] Dependent here means that they are part of a larger system and that they depend on each other to complete their function, ie, $e2$ cannot complete its work without $e1$ completing its work or visa ve rsa.

call this the law of entropy compatibility. Two systems needing to work together to produce work are more compatible (relatively more efficient working together) if their entropies are similar.

Entropy, Work and Time

How are entropy, work and time related? All else (energy, algorithm, temperature, etc.) being equal, can we measure the entropy of a system by the speed at which it does its work? Conversely, does the speed at which a system produces work output affect the system's entropy? Does order expand time and disorder shrink time?

Entropy and Work Algorithms

The algorithm is the blueprint of the steps that a system goes through in order to produce its desired work output. One algorithm may be better than another, meaning that it may produce the same work output with less energy requirements or in less time. We could say that efficiency is highly dependent on the algorithm, but that given a fixed algorithm and an idealized system, energy, time and entropy determine work output.

If an algorithm is designed to handle a system with a particular entropy and you reduce the entropy of the system further, then the algorithm may no longer function correctly, because it is designed to only function at a particular entropy level. Yet for many algorithms, less entropy will result in more work output. Algorithms of this type are preferable because their flexibility provides efficiencies to be realized by lowering the entropy of the production system. In rigid terms, a viable system produces work when energy is applied over time. The type of work depends on the system algorithm. For a specific entropy, the work output will be proportional to the energy applied over time.

We can introduce the concept of the applicability of an algorithm as it relates to the order of a system, i.e., the generalness of the algorithm, which will be abbreviated as Ag. We can describe this concept by saying that the greater the Ag of an algorithm, the greater is the band of entropy in which the algorithm works for a particular system or

$$Ag => E2 - E1 \text{ or } Ag = k \ (E2-E1).$$

Another way to express this is that Delta E/Ag = constant k. If this is shown to be true for all systems, or at least for a set of systems, then k becomes sort of a universal constant, say Kg, which allows us to determine the generality of an algorithm by measuring the entropy band in which the algorithm functions, or conversely, we can know the entropy band by measuring the generalness of the algorithm. There is another relationship that we may be able to derive from these considerations, and that is the relationship between the workable entropy band and the band of energy required to make a system do work. As the entropy decreases for a given workable algorithm, it is expected that the amount of energy or time needed to do the work will also decrease. For very large entropy bands, there should be a corresponding large band of workable energy, for narrow entropy bands, a narrower energy band; and likewise for time. Therefore the greater Ag, the greater is the workable energy or time bandwidths. Is the energy band or time band divided into the entropy band also a constant for a particular Ag?

Observations on Energy and Time in a Subsystem

Let us consider time and effort where "effort" equates to expenditure of energy, either mental, physical or both in a system, in order to do work, and time expenditure as "relevancy". We can think of relevancy as the usefulness of accomplishing a work unit within a defined timeframe. Together, time and effort help determine productivity. Productivity (P) is defined here as the amount of work (W) completed divided by the time (T) and the effort (Ef), or

$$P = k_p * W \ / \ T * Ef, \text{ where } "k_p" \text{ is a productivity constant.}$$

The dictionary example used previously illustrates the time relevancy of work, which is, that the usefulness of the work has a time dependency.

Relevant work has more value than work completed outside a useful timeframe. Relevancy also depends on the size of the timeframe. The timeframe is particular to the work, the receiver, and the framework of the receiver. As a work timeframe expands, relevancy decreases and entropy increases. As a work timeframe narrows, its entropy decreases, and relevancy increases. The change in entropy within the timeframe is based on the probabilities for the work being performed at a specific time within the timeframe. The higher the probability of relevant work, the lower is the entropy of the work timeframe.

Entropy Measures Through Error

In this section we turn our attention to work performed by information systems. Let's consider systems where the work output is the structuring of information, i.e., making more order in a set of information. Now let us associate entropy and information error. Error is the unintended result from a system receiving an intended input. Simple systems are constructs that have limited degrees of freedom or behaviors. A simple system may have only so many paths of operation or outputs for a given input. Of course, one of the outputs is always the null set, or no output. If the operation of a simple system can be mapped, then each input can be expected to yield a determinate number of outputs. What happens if the entropy of a simple system is increased by adding degrees of freedom? Then the mapping will change because new functional paths have developed within the system.

The more ordered a system is the better it can be described. The better it can be described, the more predictable will be the outputs or, in other words, the probability will be higher for an expected output. The more predictable are the outputs, the lower is the chance that an input will result in an unintended output or error. With this simple reasoning, we hypothesize that as the entropy of a system increases, the possibility of error from the system also increases.

Regression analysis serves as a predictor of outcomes based on historical or prior input/output relationships. Increasing the number of independent variables that strongly correlate with dependent variables decreases the error of the regression analysis. Or more strongly correlating existing independent variables with the dependent variable will decrease the error of the prediction. Each of the independent variables that affect the outcome of the system has an error in predicting the outcome of the system. This error could be equated to the entropy of the system. This is stated as a formula, where k_r is a proportionality constant for regression analysis, as:

$$\text{Regression Analysis Error} = k_r * \text{System Entropy}$$

Consequently, if we can determine the regression analysis error of an information or other bounded system, we will know the entropy of the system. By measuring the error of a prediction (based on history), we have measured the entropy of the system. The importance of this relationship cannot be underestimated. It gives us the ability to determine the entropy of a system free of the energy-work relationship, which is often difficult to measure.

Entropy & Redundancy

In information theory, adding information to secure the transmission of a message also increases the entropy of the message. This added information can be a repetition of the entire message or more cleverly a smaller subset with an algorithm that ensures correction of the message should it get corrupted during transmission. In either case, the additional information adds to the redundancy of the message and increases its entropy.

One can also think of redundancy as linked with degrees of freedom. If I have one path from point A to point B, then there is only one way to reach point B from point A. But if three paths link the two points, then I have redundancy in the paths and also additional degrees of freedom if the function in question is to get from point A to point B. With more paths, one can choose the route from point A to point B and consequently, there is more system entropy since at any time during the travel one could be on at least three different paths.

If we agree that entropy and redundancy are related proportionately, is it also reasonable to assume that this relationship is stepwise? As redundancy increases, entropy increases. When redundancy increases entropy sufficiently, redundancy may begin to be recursive, for example, duplicating itself. When this happens a "shiftpoint" will occur where the entropy can suddenly increase stepwise. At this shiftpoint, entropy takes a large jump before once again varying smoothly. At this shiftpoint, system efficiency may also take a larger than expected drop. Is it reasonable to assume that at this shiftpoint, a bounded system has gained a degree of freedom?

Principles of Entropy and Work in Systems

From the discussion given in the above sections, we see that energy, work, entropy, time, efficiency, and redundancy are interrelated in the functioning of systems, and that it would be reasonable to formulate some basic hypotheses for these relationships, which are listed in Appendix A.

~~~

Chapter III
The Entropy Spectrum

Systems have entropy and they also have an entropy equilibrium about which system entropy varies. It is this equilibrium that helps define a system's properties. The entropy of a system may vary above and below this equilibrium point, but the entropy can only stray so far from its equilibrium point without the system losing its properties.

As an example, think of a rubber band at rest versus a rubber band that has been stretched. They both have the same system properties but their entropies are different. However, if the rubber band is stretched too far, it loses some of its system properties, namely its original structure.

Let us then consider that just as electromagnetic phenomena are spread on a spectrum using wavelength or wave frequency as the spectrum increment, why can't we also diagram system entropies on a spectrum, where the amount of entropy is the spectrum increment. We can call this the "Entropy Spectrum" with the left side of the spectrum having the lowest entropy and the right side the highest entropy.

If we limit the systems measured to electromagnetic phenomena, then this portion of the entropy spectrum would display electromagnetic phenomena using the disorder of the phenomena as the measuring increment, rather than using their frequency as in the electromagnetic spectrum.

The entropy spectrum for a series of similar systems is a continuous band but with a number of equilibriums in that band. For many elements and compounds, there are three distinct entropy equilibriums, one for each of the physical states of matter - gas, liquid, and solid. So for example, the entropy spectrum for water would have on the left with the least disorder a crystal of ice, in the middle with higher disorder, liquid water, and on the right with the most disorder, steam. Each of these states of water would have one equilibrium point on this spectrum for each phase of matter. The full entropy spectrum would include all the elements and compounds in nature and the example given above is of just one compound of matter.

If we take a much wider view of this conjectured entropy spectrum, we then arrive at a spectrum that contains all known matter and systems arranged according to their entropy equilibriums, with for the sake of consistency, the left side of the spectrum being the most ordered and the right side being the most disordered. At the leftmost side of the spectrum we have an endpoint that is a perfectly ordered and a system that is an unobtainable construct with zero entropy. At the rightmost side of the spectrum, we have an endpoint that is a construct with infinite entropy.

As an example of a piece of this spectrum, we can select the following systems: a balloon filled with room temperature air, a diamond, an ice cube, a pencil, and an effluent of steam. On the entropy spectrum, these systems would be arranged from left (low entropy) to right (high entropy) as follows:

Diamond — Ice Cube — Pencil — Balloon — Steam

It is clear from our look at the states of water that temperature has a clear effect on the entropy of a system, in this case a group of water molecules. As the temperature of the water increases, the entropy of the water increases. So while water, or H_2O, occupies perhaps a single point on the macro entropy spectrum, it also occupies a micro-band on the spectrum with three equilibrium points. We could conjecture that the same is true for electromagnetic phenomena. While light may have a small area on the macro entropy spectrum, we know that it exists as a band of different colors of light and other forms of radiation when looked at more closely, each form with a different entropy equilibrium.

While the electromagnetic spectrum classifies electromagnetic phenomena according to their wavelengths, the entropy spectrum can classify all systems that can be identified and whose entropy can be measured. The entropy spectrum can provide a new way of looking at the universe and everything that it contains. A broad view of this spectrum would have matter near absolute zero at one end and large gaseous clouds and radiation at the other. Heat, light, and other types of radiation would all have places in this spectrum. Just as the electromagnetic spectrum was theorized and then measured, a challenge to scientists is made here to develop a comprehensive Entropy Spectrum for the universe, with entropy equilibriums for major measurable systems and phenomena.

~~~

Chapter IV

Entropy and Creation

What might happen if the entropy of a system should ever reach zero? A mathematician friend suggested to me that "the system would explode." I have considered his response many times, and it makes perfect sense.

A system that gets lower in entropy must become more rigid. It must have fewer and fewer degrees of freedom. As the entropy becomes less, the system goes through many changes, and it is likely no longer recognizable as the original system. As degrees of freedom disappear, the system not only condenses but it also gets colder. If the temperature of a system is related to free movement of molecules or atomic particles, then this movement carries with it an amount of entropy. At near zero entropy, the system becomes very compact and very cold (near absolute zero Kelvin). If a system were to reach zero entropy, all molecular or atomic movement would have to cease. In this condition, atomic particles would no longer be reacting or moving in response to nuclear forces, and the forces of gravitation and magnetism. Imagining such a system involves analysis of many nuclear component theories developed in recent years. From a conceptual perspective, would a system with zero entropy continue to exist in physical space, or would it disappear from physical space yet continue to exist, or would it simply explode to regain some entropy? If we choose the latter conjecture, then it perhaps explains what happened at the creation of the Universe.

Continuing the previous conjecture, at creation a system was so completely ordered, so perfect, it exploded into being. But what would a zero-entropy system be like? Would it be matter, matter and energy, just energy, or none of these? Could the system be so ordered so as to be simply the *perfect thought* of God, the Creator?

If a system of matter transitions so that entropy decreases continuously, its electrons must start to slow down, because, in motion they represent uncertainty, and uncertainty is a form of disorder. As long as electron position and energy or momentum are probabilities, then the system cannot be perfectly described, and hence it must have some entropy. As entropy decreases, probabilities are more certain or higher, because the system becomes more rigid and has less degrees of freedom. We could say that as entropy approaches zero, probability approaches perfect predictability[13], or that as

$$S \longrightarrow 0, P \longrightarrow 1.$$

The mathematics is in an inverse relation. The explanation would say that as entropy diminishes in a system, the system becomes more determinate and rigid. In addition, as a system becomes more determinate and the probability approaches unity, the usefulness of the probability diminishes. Knowing the probability is less important in a very low entropy system, because certainty is higher. When the entropy becomes zero, the probability exists only as certainty.

Quantum Theory (QT) stipulates that the position and energy of atomic particles cannot both be determinate at the same point in time. In order to preserve this central concept of QT, we could agree that as position becomes more determinable, energy becomes less determinable.[14] Since QT prescribes a non-deterministic world, as the entropy of a system decreases and position becomes more deterministic, then energy adopts more extreme values. In a near-zero state of entropy, one possibility is that

---

[13] The importance of probability diminishes, because at very high probabilities, certainty is very high. The knowledge mass or information value of the probability becomes less and less, approaching zero.

[14] If energy is less determinable, it implies that it has a wider range of values between a theoretical high and low, which implies that the energy could with high probability be very low or very high.

matter becomes so deterministic and insignificant in comparison to the bound energy, that the system transitions to a system of physical energy. For the system to remain one of matter and energy at near-zero entropy would imply a deterministic state that violates the rules of QT.

Accepting the above possibilities, we conclude that when the entropy of a system of matter reaches an extremely low point, the matter of the system essentially transforms to energy. If the entropy of light were to be increased through experimental means, then we theorize that light's behavior would be more like a particle with mass than a wave of energy. In short, entropy adds structure and mass to energy. Without entropy, structure is reduced to a point and mass is condensed accordingly.

Extending the above conjectures, if a system of energy (say a form of electromagnetic radiation) goes through an entropy reduction process to near zero, would the electromagnetic radiation stop flowing? What would be a system of electromagnetic energy with near zero or zero entropy? First, it would likely cease to exhibit the nature of energy based on our previous assumptions about entropy bands. Second, at zero entropy, it would not be a part of a 3-dimensional world, because matter has collapsed to energy and energy has collapsed to something else.

What would that something else be? Perhaps that something else is pure thought, so pure and perfect that it has no disorder. Once conceived and released into a universe with entropy, pure thought instantaneously starts unraveling under the 3-dimensional structure with entropy in the real world. As a wound up spring that needs to unwind, it is drawn into a space that it must fill. When it starts to fill that void and acquires entropy, it transforms itself into energy, and then from energy it transforms itself into matter. Time is the unraveling of creation.

Following this reasoning, perhaps pure thought with no entropy is what created the universe, the beginning of our reality, the "big bang" from which scientists say all energy and matter emerged and flows. If we can show this conjecture to be true, then we must also accept one Supreme

Being that possessed the pure thought that created the universe and the reality we live in.

An additional conjecture: if you turn up energy on a photon (energy with no mass), it becomes an electron (energy with mass). Dirac theorized that charges throughout the universe must be balanced, so that matter must have corresponding anti-matter to balance charges. This led to the discovery of the positron to counteract the charges of the electron; likewise, proton and anti-proton. But for the physical universe to exist, asymmetry is necessary in the matter/anti-matter concentrations. This asymmetry does exist and would not be possible without disorder or entropy. It is the entropy of the universe that makes the physical universe possible. Without it, would we not have matter and therefore no physical manifestation? Without the presence of entropy, would we all exist in a state of energy?

~~~

Chapter V

Entropy, Evolution and Life

It is scientifically and commonly understood that an inanimate system left in its environment will eventually disintegrate to an unrecognizable form. It will go from a more structured and integral system of lower entropy to a more unstructured system of higher entropy. When its entropy passes through and beyond its entropy equilibrium band, the system will no longer display its original properties. Polished metal eventually turns into a pile of rust that no longer shines, millions of sea shells turn to sand particles which have none of the physical properties of those structured sea shells, rocks become pebbles and pebbles become sand, and many large systems just disappear over time into molecular and atomic components too small to see. Even at an atomic level, some elements change over time to other elements, as in nuclear decay. The evolution of matter is directed towards an increase in entropy.

It is this singular fact of science and nature, that entropy goes on increasing and that inanimate compounds tend towards smaller and smaller building blocks, as their structure unravels. This tendency and process is restricted or reversed by what we call "life" but in the inanimate or "lifeless" realm, the process of unraveling and toward higher entropy is factually recognized and understood. The facts of this process stand firmly in the way of an evolutionary theory for the creation of life through random mutations.

The scientific theory of evolution describes the beginnings and apparent anthropomorphic forward progress of life on earth, starting with random mixtures of chemical molecules organizing themselves into simple lifelike forms, which at some point acquire metabolism and begin to evolve into more complex forms on their way through the species chain of evolution. It is not being argued here that evolution does not occur; it is being argued that physical systems tend to devolve rather than evolve, and that life can inhibit devolution.[15]

Inanimate and Animate Systems

If you consider the behavior over time of inanimate and animate systems, you begin to ask yourself the question: what makes them different? Observationally, it becomes clear that the distinguishing feature between inanimate and animate systems is that the latter can reverse, delay or slow down the progress of entropy over time. As long as a system is animate, then it is able to maintain the entropy equilibrium upon which its living natural properties are based[16]. When a system looses "life", i.e., by definition becomes inanimate, it is no longer able to maintain its entropy equilibrium and it more rapidly succumbs to the progress of entropy. Inanimate systems cannot maintain their entropy equilibrium through the means of metabolism or other energy translations we understand to be "life".

Simple animate systems, such as flatworms, make up some of the least complex forms of life. It is expected that given sufficient time, these simple organisms will evolve into more complex ones by mutation and natural selection. Philosophically, we should ask whether they want to become more complex or if there exists any natural phenomenon urging greater complexity. Is there an advantage to complexity? Do not lower complexity life-forms far outnumber life-forms of higher complexity? The modern theory of evolution explains the drive to greater complexity as random mutations in the flatworms which will allow certain of them to better survive in their environment, and these slightly different (and apparently

---

[15] We can coarsely ask: does *time x chance x fitness = life?*

[16] Could we use this as the definition of life?

more complex) flatworms will eventually outnumber the more primitive type of flatworms, resulting in the dominance of a new kind of flatworm. It is very hard to empirically demonstrate a shift to greater complexity.

For progressive complexity to occur, should we assume then that these simple animate systems can evolve into more complex systems as evolutionary theory presupposes? Or should we assume that the end result of this evolution is a slightly different organism that is better adapted to its environment but still a simple flatworm that is not more complex? We are faced with two fundamental questions. First, will a flatworm ever become more than just a flatworm? Second, if a simple animate system like a flatworm, could become a more complex system with the passage of time, what compels it to evolve to a higher level of complexity? In other words, is nature as dictated by physical law, attempting to make organisms more complex, attempting to make them less complex or is nature maintaining homeostasis? Are there good arguments to support any of these assumptions?

The natural world evolves over time, and plants and animals adapt to their surroundings, but are there concrete observations that show the passage of time as being the mechanism for changing one living organism to another organism of significantly higher complexity? We can see evidence of transition everywhere in nature, but we don't see evidence that supports a scientific rationale for migration to higher complexity. In contrast, there is a sound scientific rationale, namely the effects of entropy in systems, for showing that complex organisms devolve instead of evolve over time.

In inanimate systems, it is easy to see that system entropy increases over time, from less to more. In animate systems, we may also observe the same phenomenon if energy in the form of nutrients are not available. If nutrients are available, we see entropy equilibrium maintenance as one of the main purposes of the life function. The animate system is intent on maintaining its own nature as one of its reasons for living.

As an inanimate system example, a rock made of sandstone has certain properties that define it as a rock, and dry mud has other properties that define it as dry mud, but both are made of essentially the same substances and exist in the same environment. What differentiates the two is their entropy and their different entropy equilibriums. Both sandstone rock and mud are in the same physical phase but are different in appearance and properties because the rock has greater structure (less entropy) than the mud.

A similar observation can be said for ice, water, and steam. When the amount of entropy in this system goes beyond the entropy equilibrium band that defines the system, the properties of the system change and a new entropy equilibrium is reached. The system has new physical properties and a new definition or "nature". Thus the molecule which is comprised of two hydrogen atoms and one oxygen atom has three distinct natures, or physical states and three corresponding entropy equilibriums. At each entropy equilibrium, H_2O is a different physical system. It is the absolute amount of entropy in the system that determines its properties. Energy applied to the system can change its entropy equilibrium resulting in new physical properties. Energy in the form of heat applied to ice changes the ice into water, and with further application, changes the water into steam.

In animate objects, the same is true, but with one important difference. Animate objects have the ability to adjust their entropy equilibrium through metabolism. Their entropy change is not wholly dependent on their external environment, as in the case of water. Maintaining entropy maintains the animate system's nature. So for instance, if a seal living in the artic eats enough fish, its metabolism is able to generate enough heat to maintain its nature, in spite of the extreme cold of the external environment. However, if the seal stops eating, its metabolism will slow down and its internal entropy will change until it ceases to be alive. If the entropy of an animate system increases or decreases sufficiently to depart from its entropy equilibrium band, it cannot return to its previous entropy equilibrium, the way for instance, ice can become water and then ice again. While ice can change to water and back to ice with the addition or removal of heat energy, animate objects do not exhibit the same reversibility for major entropy changes.

The observation then is that when an animate system departs from its equilibrium band, then it ceases to be animate, and it is not typically re-animated by entropy change. Objects that are animate, alive, maintain their entropy equilibrium through the process of being animate. Inanimate objects keep their equilibrium solely through their atomic and molecular structure, which is at the mercy of their external environment and their internal atomic forces.

When an animate object's entropy lies within the object's entropy band, then its animate function varies inversely with its entropy. But when the object's entropy goes outside its entropy band, the animate function ceases, while the physical structure continues to change with changes in entropy. In essence, the animate object becomes inanimate in the process of departing its entropy band. It does not transmute to another animate object, but instead becomes an inanimate object fully subject to its environment.

If we assume these observations of animate objects to be true, then we should also consider the phenomenon of metamorphosis: for example, the transformation of a caterpillar to a butterfly or the tadpole to a frog? Are these transformations in the physical properties of animate objects a a departure from an entropy band? If we could measure the entropy of the caterpillar and the butterfly, and find that they are essentially the same, then this would support the above discussion negating transmutation. If the entropies of these animate systems are very different, that is, existing in different entropy bands, then the argument against transmutation falls apart and we have to re-evaluate the conjecture.

If we assume transmutation is prevented when entropy bands change, we should investigate the entropies in the lifecycles of different parasites, such as in malaria. Can we interrupt the malaria parasite lifecycle by sufficiently changing the entropy of the organism in one of its lifecycle stages?

On what does the entropy equilibrium of an animate system depend? At a minimum let's consider a simple system like the flatworm. First, this system has structure which affects the base entropy. The next major component is the system algorithm for the metabolism of the flatworm which creates a particular entropy. Finally, you have system deviations from the algorithm caused by external or internal energies. The temperature of the medium that surrounds a system is one example of an external factor affecting heat transfer in and out of the flatworm. The system structure and functioning algorithm(s) contribute to the system entropy. The entropy equilibrium is the optimum entropy for the proper nature of the system and lies within a continuous band of entropies. The system entropy can vary within this band of entropies. When the system entropy moves outside the band, i.e., beyond the entropy band limits, then the nature of the system must change.

If one accepts this explanation, a question that arises is how difficult is it to jump from one entropy equilibrium band to a new one? It can be argued that for life forms, departing an entropy equilibrium band is equivalent to the departure of life from its form. This then presents scientific and philosophical considerations of the purpose of entropy and entropy equilibriums: is entropy integral to the requirements for life and do entropy bands act as a barricade against special evolution?

The best evidence we have for supporting these conjectures is that we observe that organisms don't go through special evolution in the face of catastrophic environmental stressors. Instead, organisms die when the entropy of the organism is forced outside its entropy band by environmental factors.

We could conclude that maintaining entropy equilibrium is central to maintaining life. This is an argument against any misconception that life forms evolve into different life forms during their lifetime. Evolutionary science today rests primarily on the theory of natural selection and the drivers for variation in natural selection are random mutation and many progeny. While scientifically sound, this mechanism requires very large timeframes for species to evolve into different species.

Adaptations occur and genetic mutations occur, but the life form does not depart from its entropy equilibrium band. If it does, it ceases to have life, and so it either retains its nature or it is erased from the spectrum of living organisms.

On the other hand, intelligent design of plants and animals through selective and forced breeding does yield new forms with different entropy equilibriums at much faster rates than dependence on random genetic mutations. These new forms are departures from previous natures but not new species.

Human Control of Entropy

Human beings are different from other life-forms in the degree to which we set about to order and control our environment. We not only use our minds and physical attributes to order our surroundings to minimize environmental entropy, but we also create tools (an act in itself of reducing entropy) to help us order our environment. The use of tools helps us reduce much more environmental entropy in finite increments of time and energy than would be possible without tools. In our modern era, the tools we use are very advanced and benefit from machinery that convert energy into work output. The idea of using nature's energy supply to replace human mental and physical exertion through development of sophisticated machinery is relatively new in the span of human history and mainly segregated to humans and not to other animals. Prior machinery replaced mostly physical exertion[17] but this machinery was energized by human or animal exertion, not by fossil fuels as is the case today.

For humans, the creation of tools is a step-wise process. That is, tools are perfected over many years but the process often exhibits non-periodic radical improvements due to invention. Although small continuous improvements occur in some areas, such as cooking, we remember historically mainly radical changes in tool refinement or re-definement that advance civilization, or man's ability to structure his environment.

---

[17] The abacus and astrolabe are two examples of ancient machines to aid mental exertion. The computer is the revolutionary machine for current times that alleviates and expands human mental exertion.

Some animals other than humans exhibit the creation and use of tools, namely in the primate species, but the step-wise process of refinement appears to be missing, most likely because the transfer of generational knowledge is much more limited than in humans. By codifying knowledge in a way that it can be passed to future generations, the human species has provided the means for asynchronous inter-generational tool refinement. The sophistication of a species' communication, especially codification, provides the ability to store and communicate knowledge between generations, and this affects the development and sophistication of tools within a species and the ability to control their environment.

If we think of environmental entropy reduction through the use of tools, the more effective the tools, the greater the reductions in entropy for a given time increment and amount of energy. There are three ratios that we can measure: the time expended with a tool divided by its entropy reduction, the energy expended with a tool divided by its entropy reduction, and the environmental entropy increase associated with a tool divided by its localized entropy reduction. These entropy reduction ratios remain fairly constant for a given work algorithm and a given set of tools. When a new set of better tools is developed, the entropy reduction ratios undergo significant change. An examination of history would look for these step-wise changes and derive from them some interesting conclusions about what to expect in the future.

Besides humans, a few animals are known to fashion some very limited tool sets to assist in decreasing their environmental entropy. Examples exist for chimpanzees, apes, and bonobos. Beavers build water lodges and dams to reduce the entropy of their environment. But for the most part, man is unique in his superior development of tools and his ability to pass on those tools, or their designs, to future generations. So while an animal may by trial and error or by accident "discover" a tool, the knowledge of that tool and how it came to be made will most likely perish with the animal unless it is taught by example, to the young generation of animals.

For instance, chimpanzees have been observed teaching their young how to use a stick to remove termites from a nest to more easily access this food source. Other chimpanzees have been observed hunting with weapons such as wood sticks, and this behavior is learned through observation and imitation, but evidence suggests is not taught. If the use of a tool is not instinctual or observed by the young, then most likely its use and benefit will not be transmitted to the next generation. Primates are great observers and we know that they mimic the behavior of other primates. Not much of their acquired observational behavior is improved. If it was, their observations during interactions with man would give their offspring a significant and increasingly more sophisticated facility with tools, which evidence suggests does not occur from parent to off-spring. In this consideration of knowledge transfer, we have not addressed necessity. A worthwhile experiment to conduct in knowledge transfer would be to make a tool necessary for survival, and then see if chimpanzee adults transfer the knowledge of the tool to their off-spring.

With man, the progress of one generation in controlling environmental entropy is passed on to the next generation by example, but also through codified knowledge in societies that progress. In this way, the progress achieved transcends a single person's life and becomes cumulative as new generations improve on what has been codified. The transfer of ideas through codified knowledge allows for development of tools to span many generations, hence the great value of education and the benefits of a structured society that permit the education and preservation of codified knowledge.

For example, consider the development of a tool known to us as the submarine. As first conceived and codified in manuscripts by Leonardo Da Vinci, it was not very usable, but the codified knowledge served as a basis and inspiration for future development. The original conceptual design was refined and re-designed in future generations until a usable version was developed. The principles first illustrated by Leonardo found their way into a usable model generations later. Further development of the submarine came in the form of more codified ideas, proposed in the novels of Jules Verne as science fiction. These ideas in large part became reality as they were the inspiration in the 20th century for development of the modern submarine.

Man's ability to intellectualize and codify the process of experience-based learning and thus preserve that knowledge is what allows humans to tame the environment at a much faster rate and more effectively than other animals can. Again, the end goal of man's constructive efforts very often involves decreasing the entropy of the environment to make it safer, more comfortable, and able to provide sufficient food for the species. But sadly, sometimes man's over-control of the environment can serve as a means and justification for environmental destruction and the misuse of nature's gifts to the species. This destruction can interrupt the delicate balance that nature maintains, which in the terms of this paper is to maintain its own entropy equilibrium. For life to continue on earth, the super-system we call nature must balance order and disorder. Global temperature is just one indicator and contributor to this balance. The others should be identified, along with their effect on the entropy of nature.

As better tools are developed and as knowledge is accumulated, man is more able and more efficient at decreasing the entropy of his environment. For example, man's current accumulated codified knowledge of the strength and fabrication of steel and other invented alloys allows all kinds of entropy-defying structures to be built by the human species. If man were to lose this knowledge of metal alloys and their properties, such as the tensile strength measurements of different alloys at different temperatures, then the amazing structures we build so quickly today would no longer be possible in a single generational timeframe. If tools or the knowledge of some tools are lost, man's control of entropy can be considerably diminished. For example, the ancient building techniques of the Egyptians and Greeks are difficult to discern or duplicate today.

From the above examples, we might assume that all the organisms on earth desire to be in the process of reducing environmental entropy, but that assumption is not true. There are organisms on the earth that are in the process of increasing entropy. They are taking what human energy has ordered and are breaking it down back into the previously disordered constituents. A good example is the action of bacteria, which are in the process of breaking more ordered compounds into less ordered compounds. Whereas plants take nutrients and water from soil and convert them to sugars through photosynthesis with energy from light (an entropy reducing process), bacteria are busy converting the cellulose and sugars of plants back into the basic nutrients in the soil. Furthermore, other types of life which are pathogenic to animals act to increase entropy by killing more ordered life forms.

Human beings are also in the process of increasing entropy, because they choose to or because they do so unknowingly. Examples include when human beings engage in war, where energy is used to break down constructs of order previously created by humans and to kill human life, which removes some of the human potential for entropy reduction. In fact, war dissolves social constructs set up to order life by rendering them inconsequential when life itself is at stake. War creates famines bby disrupting food supplies thereby relegating humans to spending their time foraging for food. Maslow's hierarchy of needs quickly drops to its lower levels in the face of threats to life and reproduction. When this happens, the need to decrease entropy by refinement is subjugated to the need of survival. What we consider to be "societal progress"[18] often comes to a crawl during periods of conflict. In a peaceful setting humans resort to building lasting reductions in entropy. More importantly, fostering a structured society is highly desirable so that thought applies virtual order to physical disorder in the present and is passed into the future as actionable knowledge.

Is there a battle going on in the human species between entropy reducers and entropy producers? It certainly seems that way. Free will allows human beings to expend their energy increasing entropy or decreasing their personal entropy or the entropy of their surroundings. The reasons behind these actions are unclear but are part of the human condition. Should we then consider the philosophical or religious basis for entropy reducers

---

[18] An examination of history can show that a great deal of invention and refinement occurs during periods of conflict. But what part of these should be considered human progress. Is the ability to kill more people more quickly a sign of progress? Also, history cannot show what new progress would have been made in the absence of conflict.

45

versus entropy producers in the human species, where choice and free will are considered part of our nature?

Some organisms spend their life increasing the entropy of their surroundings while others spend their life reducing it. Are both processes important? Is one process good and one bad? Is the overall balance of entropy in the closed system of our planet earth increasing or decreasing, or does it remain constant? Are animate systems necessary to reduce environmental entropy? These are questions that have spiritual, philosophical, and scientific arguments. How do we answer such questions, and what do the answers tell us about life and our purpose on the planet?

Evolution or Devolution

Since the dawn of Darwinian thought on the evolution of life through natural selection, the direction assumed for this evolution has been from systems of less complexity to systems of more complexity, and apparently resulting in homo sapiens. According to what is commonly called "evolution", the one-celled life form eventually works its way up the evolutionary ladder to the complex invertebrate and then to the vertebrate. At each step, greater specialization appears in the biology of the life forms. Skeletal frameworks are organized, the nervous systems become more centralized, internal organs differentiate and become more specialized and numerous, and intelligence becomes apparently more concentrated in one nervous system location and becomes apparently more sophisticated. This is the evolutionary ladder as it is currently understood and taught.

As we move through evolutionary history, or up the so-called "evolutionary ladder", life forms evolve into more specialized and complex designs. The idea that life forms might devolve instead of evolve would be problematic to the apparent progression needed for evolutionary theory to account for complex life forms.

For example, the common course of scientists exploring the development of man is to assume the evolution of the monkey to the ape to the humanoid. There are numerous books[19] that describe the painstaking efforts being made through DNA analysis to discover the links, timelines and processes for supporting this evolutionary ladder between the members of the primate species. One major issue with such theory is the direction of the evolution from more primitive to more developed primates. In general terms, evolution from less complex to more complex organisms is fundamentally based on natural selection, and the belief that more complex organisms are better able to survive their environment than less complex organisms. But is this idea based in scientific fact? We could argue that the forward march of complexity is contrary to the scientific facts of entropy, without some outside force beyond the environment directing or encouraging the complexity.

In the current scientific development of primate evolution, researchers are looking for a number of linkages. One linkage mechanism involves hybridization or the cross-mating of species in order too change the gene pool and provide the bridge between primates. But as scientists empirically point out, most hybrid animals do not reproduce well or not at all, and eventually a hybrid strain declines and disappears. This supports the ideas in previous chapters of each system having an entropy equilibrium at which its unique nature is most strongly expressed.

To scientifically support hybridization within the primates, a lot of analysis has gone into the genomes of humans and apes, since tools are now available to map these out. The DNA of a Chimpanzee contains 48 chromosomes while the DNA of humans contains only 46 chromosomes. Scientific theories are being tested for how primates moved from 48 to 46 chromosomes via hybridization as an evolutionary path to humans. But not discussed in the literature is the consideration that the primates could have moved from 46 to 48 chromosomes.

Scientists have to ask themselves if it is more *scientifically* defensible to consider whether humans evolved from chimpanzees, or whether chimpanzees evolved from humans. Developing an evolutionary ladder

---

[19] *Almost Chimpanzee* by Jon Cohen, is a notable example and a good compendium of current research.

within the primates from humans to chimpanzees may be scientifically easier to defend as more in accordance with the principles of entropy. Let us examine one such hypothetical scenario.

It is often the case that a Down Syndrome child is born with one extra chromosome. In fact, it is one of the pre-birth tests performed to identify the possibility of having such a special child. We know from current empirical evidence that one of the causes of Down's Syndrome is environmental, that is, exposure to certain chemicals before or during pregnancy[20] in females or exposure to external radiation[21] in males. Let us assume that at some point in the past, the human population was exposed to poisonous chemicals from a volcanic eruption or to radiation from some planetary activity. Exposure may have occurred from an asteroid hitting the earth, polluting the rivers and streams humans were drinking from. Whatever the source, we can scientifically reason that this exposure resulted in a sudden increase in the birth of Down's Syndrome babies (DSB). Most of these DSBs would have been born with one more chromosome than their parents. If as adults, the DSBs intermarried in sufficient numbers, the result could be a group of offspring with 48 chromosomes in their DNA, the same as what we find in chimpanzees and other ape species today.

Humanoids with 48 chromosome would be different enough to be separated from their progenitors by wandering away in groups on their own, but more likely, they would have been forced out from their human tribe because of their differences. Once separated geographically, they would have to rely on themselves to develop their own sources of nutrition, survival, and reproduction.

A reasonable scenario is that without the protection of their progenitor tribe, they would have been forced to climb trees to avoid predatory animals. Once in the trees, their nutrition would have come from fruits,

---

[20] Antidepressant use in mothers has been singled out as a causal effect of Down's Syndrome. Lawsuits are pending against manufacturers. There is a link between chemical ingestion and Down's births.

[21] In the early days of microwave radar, a causal link was established between exposure of radar technicians to microwaves and their children being born with Down's Syndrome.

leafs, twigs, insects and nuts and their digestive systems would have to change and adjust to this diet. By growing up in the canopy of the forest, they would develop their arms, hands, legs and feet differently to keep from falling to the ground. Instead of running across savannahs in search of animals to eat, they would instead move from branch to branch in the canopy while animals on the ground sought to eat them. Additionally, by spending a good deal of time crouching in the tree limbs, their upright posture would eventually erode. The exposure to the weather in the canopy would also create the need for stronger and longer body hair. Their offspring would develop even more fully these characteristics needed to survive in the outdoors without hard shelter and fire.

The vegetarian and insect diet of the canopy would not contain the complex animal proteins ingested by their predecessor human tribe, and this lack of protein and change in stimulus could result in some atrophy of brain function. DSB children have difficulty learning to speak and it is reasonable, that left to themselves, speech would eventually take a back seat to other means of communication such as gestures and specialized noises more prescient to the survival needs of the group. Without having to pursue other animals for food (or to defend themselves from ground-based predators), the group would have fewer stimuli for developing higher levels of verbal communication[22]. It is plausible to reason then that in short order human language as we know it would disappear from the behavioral characteristics of this new group of primates.

A life of forest canopy living would allow survival from ground-based predators but would also change a human's skeletal and muscular generation. The more simple life needs of the canopy (e.g., availability of food without the need for hunting, tools, and fire) would result in the atrophy of hip and leg muscles used for running after animals, and the enhancement of hand size and strength for gripping branches. Foot size and shape would also change to accommodate gripping rather than walking or running long distances.

---

[22] The ability of a group of homo sapiens to coordinate the hunt for food or protection is considered by evolutionists to be the primary stimulus for the development of more complex verbal communication.

It is hard to know how many generations it would take for a splinter group of humanoid DSBs to acquire ape-like characteristics, but it is more scientifically correct to make the case for this progression than it is to argue that humans evolved to greater complexity from other primates through hybridization. Under the conjecture of devolution, the primates all started with language and hunting skills, but some lost these through a combination of genetic alteration and the requirements of survival as a rejected group. This example of devolution as differentiating the primates is scientifically more in line with the principles of entropy in biological systems.

Less entropy supports more complex structures with more usable energy while more entropy supports structures that are less complex with less usable energy.

~~~

Chapter VI

Entropy and Health

As biological creatures living in a physical environment, we experience and accumulate stress associated with the entropy of our external and internal environments. This stress accumulation can alter biological processes and that can cause secondary disorders. If it does, these secondary disorders change the entropy of the biological system, and if a sufficient entropy change occurs, the biological system no longer functions as intended. The entropy equilibrium of the biological system can shift, creating a modified system that will not function fully as intended, or if the shift is sufficient, can begin to function unnaturally. This is the root cause of cancer in a biological system. The stress of forces acting on the biological system as a whole or on particular subsystems can change the nature of the subsystems and therefore their function. The nature of each subsystem can remain intact if its entropy equilibrium remains within its equilibrium boundaries.

As conscious biological organisms, human beings have above all the free will, then the perception, intelligence, mobility, energy, and tools to be able to decrease the entropy of their current and future environments.[23]

---

[23] Does this make us unique from other living things? Not at all. Birds feather and build their nests to contain the young that have not yet been born. The nests represent a bird's effort to mold the future environment to better serve the needs of incubating their eggs in a safe place, and then to contain the chicks when they hatch. The nest is created by birds to order their environment for the purpose of hatching their eggs and raising their young. The birds are reducing the entropy of their natural environment for a future purpose, because their natural environment does not provide suitable safety. So, as humans, we are not unique in our desire or ability to decrease the entropy of our surroundings.

But what about the entropy of the full human being as a living biological system, in other words, a human's personal entropy? We are comfortable with decreasing the entropy of our environment, but do we act as decisively or as constructively to manage our *personal entropy*?

Many studies in health have emphasized the role of nutrition in maintaining good health. That is because certain foods contain compounds that assist the body in fighting disease[24], in maintaining a certain resistance level to the onset of disease, by providing the nutrients needed for proper cell function. Other foods deplete the body of healthy chemical compounds or create conditions that adversely affect cell function. The body and its immune system are comprised of very complex bio-chemical subsystems that involve countless interactions between living cells and their environment, in all parts of the body. Each cell or class of cells can be thought of as a closed physical subsystem for the purpose of establishing its entropy equilibrium. Knowing this equilibrium for the subsystem could be very useful.

Research has drilled deeper and deeper into the minutiae of the body system, detailing in many instances, the precise chemical and biological mechanisms and interactions that occur at the cellular and intracellular level, from the behavior of proteins as gatekeepers on a cell's surface, to the cell's internal energy functions and the genetic transfer of information.

What if we step back and consider a more macro perspective of the body as a single whole system made up of many subsystems? A healthy human being may show no outward signs of disease, but we know that the human body carries many disease causing organisms that are held in check by agents in the body. When we use the word "healthy" to describe someone, we are saying that the whole body system is in equilibrium with the pathogens that attack it and that it is able to maintain this equilibrium to its advantage. While there are many viruses and bacteria in the body all the time, the system as a whole is in balance, and that is what we refer to as healthy.

This healthy balance point is a reflection of the whole body's entropy

---

[24] Allicin in garlic, as one example, is an anti-viral.

equilibrium, which is not unlike the entropy equilibriums in other physical systems, meaning that the equilibrium exists at a point along an entropy band with upper and lower entropy boundaries. We conjecture here that the entropy band is for the whole body system and that the entropy equilibrium point is most likely unique for each individual.[25]

The human immune system is highly complex and functioning all the time. It is not a system that lays dormant and is suddenly called to action, but rather, it is a system that works continually at varying degrees of intensity to maintain health in the body. It is also a system that relies on order with preset (inherited) or conditioned (learned) patterns or algorithms for carrying out the work of the health maintenance processes in the body. Added to this are the patterns established by environmental exposure and by chemicals that are ingested accidentally, or intentionally as medications.

The body is normally able to maintain order, that is, proper function, in its many subsystems. Each subsystem has its own entropy band. If we were to measure the entropy of each subsystem, it is proposed that the proper function or health of each subsystem could be evaluated by comparing its current values against normative standard values.

If we view the immune system as a collection of subsystems that do the work to maintain the immune response, then health is directly related to the body's ability to maintain proper entropy in the subsystems it uses to fight pathogens.[26] When the body's immune system cannot maintain proper order in one or more health maintenance subsystems, then the

---

[25] How else can we explain the great diversity of resistance among individuals to diseases? If we were to let out a pathogen, such as the flu virus, into a room of 100 individuals, not all individuals would contract the flu, though all are equally exposed. Is it not reasonable to assume that resistance to contracting the flu is due not just to genetics but also to the condition of the body which is indicated by its holistic entropy?
[26] The order or entropy level is maintained with the help of numerous physical factors such as: metabolic rate, temperature, nutrition intake, physical and mental stress, pathogens, state of mind, hydration, and rest. "Rest" refers here to the physical act of placing the body in a state of repose such that the body is able to conduct its restorative functions. Restoration is an entropic function that should be fully explored.

equilibrium for the major system may be sufficiently altered such that its function is compromised and the body begins to become unhealthy.

As discussed in Chapter 2 on the production of work, if the entropy in an immune subsystem increases, the amount of health maintenance that can be produced for the same amount of energy decreases. For a given energy quantity, the immunity protection to an individual is less. Adding energy (e.g., through increased nutrition) can achieve the desired amount of health maintenance work at the higher entropy level. However, the amount of extra energy that can be applied effectively is limited by the equilibrium boundaries of the immune system(s). For this reason, adding too much energy may actually have an adverse effect by causing excess energy to increase the overall entropy of the health maintenance subsystem, affecting its ability to function.

If we accept that abundance of energy without entropy-reducing work can increase the disorder of a system instead of decreasing it, then this opens a whole area of consideration in the treatments of health. For instance, when does eating more food to increase energy levels benefit the immune system and when does it hamper it? If the energy goes more toward increasing entropy rather than increasing immunity work, then the immune system is being hampered. Also, a system that is hotter contains more energy, but is this heat energy contributing to work inside the system? Often, external heat is used as a body healing mechanism. If the applied heat energy is not useful, then it may lower immunity by adding entropy. If on the other hand the added energy is used to increase the work of any of the health maintenance processes, then the heat energy is a factor that may benefit body health.

At a macroscopic level, if higher blood temperature increases the production and distribution of antibodies or increases the entropy of pathogens, then warming the body slightly will help to shift the health maintenance function in a positive direction. Warming the body can be accomplished internally through exercise or specific nutrition, and heat can be applied to the body from the external environment. If the body temperature is already high enough for optimum immune function, adding more heat energy can only hinder the immune response.

Because the body is so complex, and so many processes are occurring, it will be necessary to identify which processes are most important to health maintenance, and then to reinforce the function of those processes by decreasing their entropies within the range of their entropy equilibrium boundaries. Doing so should decrease the entropy of the whole body system and thus increase health and vitality.

This discussion can be summarized by the following propositions:

1) "Health" is a whole body system equilibrium that depends on the entropies of the underlying subsystems of the body's health maintenance process.
2) The whole body system entropy is an indicator of the body's overall health, and if measured, can one day be used as a health indicator.
3) The level of entropy of a health maintenance subsystem determines how effective that process will be in contributing to the health maintenance process.
4) There is an entropy band below and above which a health maintenance subsystem will not function correctly. Maintaining health requires keeping entropy within the entropy equilibrium band.
5) The entropy equilibrium band of a subsystem has upper and lower thresholds that are points of discontinuity on an entropy-work curve for that subsystem. Moving entropy beyond these thresholds will cause a system to functioning improperly or to cease functioning.
6) Body systems where entropy has departed from the equilibrium band boundaries behave abnormally, the underlying cause of cancer.
7) If the abnormal entropy of a subsystem can be returned to its proper equilibrium range for that system, then healing will occur and the subsystem will begin to function normally.

Sleep and Entropy

For years, research has been conducted into the purpose of sleep and how much sleep is required for most people to stay healthy. This research attempts to sometimes answer the obvious, which is that a person needs about a quarter to a third of a day of sleep to stay healthy. Perhaps it should be recognized that one of the main functions of sleep is to allow body subsystems time to restore their entropy to the optimum equilibrium level to perform their work function.

Sleep gives the body and mind restoration time, and the main goal of restoration is to bring down the entropies of the body subsystems so that they may function optimally during the next wake cycle. If we thinking of the prior analogy of the messy desk, we understand that that to maintain the desired entropy level for the desk, some time must be set aside daily or weekly to tidy up the desk and maintain its order. The process of using the desk to accomplish work automatically raises the entropy of the desk as a byproduct of the work. The same is true for the body subsystems that perform work during the day. Some may need daily restoration; others may require it less often, and the frequency of necessity is most likely highly individualized based on levels of work, efficiency of work algorithms, and starting entropy levels.

We can observe that without adequate sleep, the body's chemical restorative processes do not function adequately, and the result is likely an increase in the macro entropy level of the whole body system, and an impediment to the health maintenance processes.

Adding energy to the body in the form of nutrition can temporarily increase the work output of a body subsystem that is sleep-deprived, but it can only partially assist the restorative process. For full restoration to occur, sleep is necessary so that entropy levels can be re-adjusted to optimum levels as work function is suspended.

Cellular Systems & Entropy Measurement

All physical body systems can be reduced to the cellular level. Each cell can be considered a subsystem of the larger system it is a part of. Each living cell is following a coded algorithm to perform its functions and the ability to perform its algorithm depends primarily upon its nutrient intake and its entropy state. The closer the cell is to its entropy equilibrium, the better will be the use of its nutrient intake and the results of its function.

Considering what was previously discussed in Chapter 2, there is no reason to conjecture otherwise than that a cell system will perform its function so long as its entropy is within the entropy band (EB) for that cell type. If however, the cell's entropy is forced outside its entropy band, then we will assume that one of two things will happen: 1) the cell will cease to function and die off, or 2) the cell will begin to function in a different way that is not in accordance with its coded algorithm. In the latter lies the beginning of what we refer to as cancer.

Now let us look at a biological system made up of many cells and let us assume based on previous conjecture that the entropy of the biological system is equivalent to the aggregate of the entropies of each of the subsystem cells. We should also assume, just as for the single cell, that the entropy of the overall system is bounded and has an entropy equilibrium that is ideal for the system. The entropy of the system will vary depending on how the individual subsystems cells are functioning. It is reasonable to assume that if a sufficient number of subsystem cells depart from their entropy bands, then the entropy of the overall system will also depart from its entropy band.

Using the above rationale we have the opportunity for new types of diagnostic methods with which to detect and measure diseases in individual body systems and for the body as a whole. By measuring a healthy cell's entropy, we can establish an acceptable entropy band for the proper functioning of that cell and use this measure to estimate the entropy band for the greater system. We can also consider the converse, and estimate a single cell's entropy band by measuring the entropy of the larger system. Similarly, we should be able to do this for body subsystems in relation to the body as a whole.

We assume that the entire body, though made up of thousands of subsystems, exhibits a range of entropies in its normal range of function. If we establish what the entropy range is for proper functioning or "health", then by measuring the body's entropy, we should be able to know when it is healthy or not. If the measurement of whole body entropy falls outside the "healthy" range, then we will know that something is amiss but we may not know in which subsystem. Further testing of individual body subsystems would have to be done to identify the subsystem(s) where the entropy levels are abnormal. In short, using a technique of biological entropy measurement should give us a very early warning system for the detection of disease and for interventions needed to correct problem areas. Measuring whole body entropy over time will also inform us on the aging process and on how effective interventions are in correcting a disease.

The measurement of entropy can be done at a macro or micro level and methods will differ for each. At the macro level, quantifying heat transfer and temperature changes provides one method. At the micro level, measuring cell output for a known quantifiable input should help derive cellular entropy levels. Culturing healthy cells and measuring their entropy should help to establish healthy baselines ranges. Applying these baselines to body system measurements should provide valuable diagnostic information on the health of living organisms. Developing methods to measure whole body, subsystem, and cellular entropy levels will greatly advance diagnostic medicine.

Furthermore, we may also consider measuring the entropy of viruses to help us understand when they become virulent, and use this information to thwart their activity.

~~~

CHAPTER VII

ENTROPY AND PRODUCTION

Entropy affects how much we are able to produce with a given amount of energy in a given system. In general, the lower the entropy within a system's functional entropy spectrum, the higher will be the system productivity. Conversely, the higher the entropy, the lower the productivity will be. Productivity can be measured by output for a given energy input, so we conjecture that lower system entropy provides greater output for a given amount of energy expensed in a system to produce work. However, it should also be considered that very low entropy in a production process may be deleterious because very high order can make a production system too rigid for its work algorithm to function efficiently if changes are needed. If a system has parts that are not contributing to the work output of the system, then those parts are contributing to the entropy of that system. A system with non-functioning parts cannot achieve its lowest functioning entropy and consequently its highest productive output. We further conjecture that if a highly ordered system has a part that becomes non-functioning, this change is more critical to optimum production than if the change occurred in a system with less order. Finally, time and energy are related by entropy.

A Partial History of Production

The industrial revolution, replaced human labor with machine labor on a large scale. While ancient civilizations used machines such as pulleys and levers to increase the mechanical advantage of human or animal exertion, it wasn't until the harnessing of steam power that human or animal exertion could be completely replaced. Since machine systems are more uniform and repeatable than human systems, the amount of work produced as a function of energy was easier to quantify and more dependable. Entropy in the production process was reduced in the shift from human to machine systems, and it was then possible for energy input to a production system to be separate from human sources. This freed humans to devote time and energy to other tasks, such as analyzing and organizing the production process to be more efficient. This organization created more structure which lowered the entropy of the work process, and the production system became more productive.

With the advent of computers as a second production revolution, the organizational entropy-reducing tasks that humans did could now be mapped as instructions that are now in large part being done by computer. In addition, computers have provided increasingly accurate control of machines that make possible robotics; computer controlled machining; and assembly line management. As with the industrial revolution, the human was once again "freed" and one more step removed from a direct production-related task to more fully manage the entropy of the production process. Humans also have to maintain the machines that produce the work output but now have more time and energy to be creative in ways to improve the work algorithms, or entire paradigms of production. The computer has also helped to lower the entropy of production processes by compressing the time required for production information to be fed back to make adjustments to the production process. Much faster than humans can process and react, computers process and use real-time information to control the production process, and develop higher level automated management of this information (expert systems). The result is that production efficiency benefits from reduced entropy and time factors, and thus less energy is required to produce the final product.

Entropy Reduction in Production

One example of the above considerations is the use of Just-In-Time (or JIT) systems. These systems allow parts to be made available just when they are needed in a production process, eliminating all the work and energy required for maintaining large inventories of parts. The faster the production process in output units per time, the greater is the size of the required inventory needed for daily production. JIT systems make available the parts for a product at the time they are needed, not before and not after, eliminating the work associated with maintaining a large inventory of parts. If the part arrives before, it must be held until needed which consumes more energy for storage and retrieval and raises costs. If the part arrives too late, the production is halted, consuming more energy from interruption of work flow and raising costs as well. The trick is to make the parts available only at the time they are needed for assembly. This is the lowest entropy situation for the assembly process in production and it requires a great deal of information processing and timeliness; in short, it requires that the entropy of the supply and information systems be very low. The achievement of this low entropy in large production processes is only made possible by computer systems, and by communication channels that can link distant geographic areas. The computer systems can manage the calculations fast enough in order to track parts and maintain the supply at the proper level required by the production system. Here once again, the issue of time appears since the speed of information processing and transfer determines the amount of entropy reduction that is possible.

If JIT systems had high entropies, then the wrong parts would show up at the right time, or the right parts at the wrong time, or the wrong parts at the wrong time, etc. Without computers, it would take an army of people to track parts, compute inventories, communicate requirements, arrange timely transports, and calculate needed statistical management information. Even with an army of people to process information, JIT systems might not be possible without computers. Computers are able to process the information needed much faster than a collective human workforce, and thereby lower the entropy of the information sufficiently to make a JIT production process viable.[27]

---

[27] Essentially, if energy expended on the JIT system is less than the energy to maintain an inventory of parts, then there is a net increase in work output for a given amount of energy. The computer functions as an entropy-reduction machine.

The availability of important information in time to for it to be useful is a powerful "virtual (non-physical) force". I use the term "information momentum (IM)" to refer to the amount and the rate of information delivery, or the amount of useful information divided by the time in which it is delivered. Computers are able to increase the IM to the point where it is possible for a JIT system to operate (see Chapter IX). For each system where it comes into play, IM must exceed a minimum value to effectively lower the entropy of the dependent system. We can call this value MIM. If IM < MIM, no entropy reduction can be attributed to the information delivery.

Entropy Reduction in Business Production

As in manufacturing, the conduct of business can be more or less productive. The computer has made it possible for business processes to be many times more productive with less expenditure of human energy[28]. Many companies are making a great deal of money by mining and managing the information entropy in business processes to make them more productive.

An example is the computer software industry products that create order in information to make it more accessible. Many of these products accomplish one key aspect: reducing the entropy of business information by creating virtual order for the information. This is accomplished through databases where rules and relationships between data help to make information easier to find and potentially more useful. These computer products have one

---

[28] The cost of replacing human management of business with computer driven processes increases production but may not necessarily be more cost-efficient. There is a very large cost associated with deploying and maintaining a computer network for business processes. It has not been shown, in many cases, to be more cost-effective than relying on human management. After all, during WWII, complex business processes were intricately managed by humans devoid of the benefit of modern computer systems; and many would argue, managed more effectively. The author believes that a pen and pad often works more effectively than a computer in handling business processes, but many have been deluded by the false promises of computing, and there are few holistic studies that measure the cost-benefit of computing compared to not adopting computing. Information Technology is very expensive to maintain and often degrades customer service.

thing in common - they reduce the amount of entropy in the systems they are targeted to order. By reducing the entropy of the information and the retrieval process, less energy and time is needed to access data.

By far, the easiest business process to target for entropy reduction is information storage and retrieval. Information has its own entropy associated with it. The higher the entropy in information, the less useful information becomes. The boom in information technology software of the last 30 years is primarily based on the inherent value of reducing information entropy. Reducing the large amount of entropy in information currently drives big profits. But if information entropy were to decrease significantly or if timeliness (IM) were to decrease, the benefit from information technology products would decrease and so would the return on investment. Fortunately for business, computers generate more information than ever available before to human society. As new information is created, it needs to be categorized and made accessible to be useful. This will continue to drive the information technology industry because the creation of new information is facilitated by the computer technology itself. There is a bountiful supply of new information, and the information system becomes self-serving. However, in areas that have already experienced significant entropy reduction without creating new information, the value of applying computer technology will significantly decrease.

An analogy would be hiring someone to organize your office. Once the office is organized, you no longer need the person if you maintain the order regularly. In the absence of a major change, maintaining the order is a simpler task requiring much less entropy than organizing the office. Once computer software lowers the entropy of information and the benefits are achieved, additional use of the software will provide marginal lowering of entropy and marginal benefit to the business process. At this point, all that is really needed is a maintenance process for incremental change.

It might be worthwhile to consider at this point the work required to maintain low physical entropy. Some systems absorb structure and maintain it; other systems absorb structure and release it. At one end of the system spectrum, for example with steel, a material retains its structure for a long period of time and releases it very slowly. On the opposite end of the system spectrum, for example with ice, its structure (or entropy equilibrium) is maintained only under constant application of energy and or work.

If the entropy of a production system becomes too high, then a good part of the energy available for work is absorbed by the disorder of the system which can include maintenance to the system to prevent further increases in entropy. Also, the production paths may be less efficient, because the entropy of each production process is generally higher and demands more work. For instance, if the production environment requires the assembly of three parts, and the production of one of the parts requires more heat because the part material is contaminated[29], then the additional heat will require more energy and possibly more time. The additional energy may require the process for one of the other parts to forfeit some energy; thereby altering that process as well. The final assembly of the three parts will rely on the path for the part with the longest production time.

Redundancy in Production

System entropy that is too low may have a deleterious effect on production. Some amount of elevated entropy provides alternatives that can be taken due to changing conditions in a system of production. If the entropy is too low, the flexibility of the production process is diminished and a change in product requirements, or some other parameter, may be more difficult to accommodate. A system where the entropy is too low has too few alternate paths of operation, if any.

To explain this conjecture further, we need only understand that redundancy contributes to entropy. In a highly ordered system, redundancy is greatly reduced, and the non-functioning of one part of the system can stop some (if not all) of the work produced by that system. In a less ordered and more redundant system, alternative paths can exist for work to be performed when a part of the system becomes non-functional[30].

---

[29] Material contamination is caused by the introduction of disorder to the material, i.e., the material's entropy equilibrium or "true nature" has been altered by the contamination.
[30] Not performing its intended function within the system's work algorithm.

Because of redundancy, the failure of one part does not affect as greatly the work output of the system. We are saying that lower entropy increases work output of a system, but under non-ideal conditions (where failure is possible), very low entropy can be a hindrance to work output when one or more parts of a system fail. Redundancy increases system entropy and decreases production per unit of energy, but it also decreases the potential for system failure. We could surmise that adding redundancy to a system widens the entropy band (EB) of the system such that the system's nature is maintained over a wider range of entropies.

While this can be considered mostly common sense, there is an important principle at work. As redundancy increases, productivity decreases, but if a system of parts loses the function of one of the parts (if we want to account for failure), then over time the productivity of the system with more redundancy will likely be higher than the productivity of the system with less redundancy. The low redundancy system over time becomes less efficient than the high redundancy system when failure probability is high. The longer the timeframe under consideration, the lower entropy system suddenly loses its productivity advantage over the higher entropy system when failures exceed relevant degrees of freedom.

If we think of redundancy in terms of "degrees of freedom", then the more degrees of freedom a system possesses, the less productive it will be than a system with less degrees of freedom. Assume we have two systems with the same work algorithm, but with different degrees of freedom. If a degree of freedom is removed from each system, the system with more degrees may become more productive than the system with fewer degrees. Take for example a silicon substrate that has been given impurities to make it semi-conductive. These impurities raise its entropy but also allow it to perform work by transferring electrons when energy is applied. There must exist a threshold above which removing degrees of freedom increases work output and below which removing degrees of freedom decreases work output. This threshold is a point of discontinuity on the entropy-work spectrum. If entropy is reduced below the threshold, then this has a deleterious effect on production. Each system has its own entropy threshold determined by the degrees of freedom in the system and the work algorithm which determines how energy is used.

Entropy in Design

The design of buildings, products, machines etc. should account for the role of entropy in the functionality and longevity of the design. When you walk into an unfamiliar building, is it obvious where to take the next step or is the design of the interior confusing the intended flow of people. Do products incorporate designs to mitigate the onslaught of entropic forces? An example to consider is outdoor solar lights that currently mostly fail within a two-year period. How does entropy in financial systems affect the reliability and speed of the information they produce?

Work that is confined to a mechanical system can be idealized to exclude human interaction. As an example, consider work done in producing electric power from water turbine generators in a hydroelectric plant. The turbines are mechanical systems that run on water pressure and do not require human interaction unless there is a maintenance need. The work produced by such a mechanical system is dependent on the kinetic energy of the water released from a height, the structural entropy of the mechanical system, and the friction of the mechanical system. We may also consider that friction is the result of a specific form of structural entropy. The fact that energy transfer carries with it an additional entropy gain to the environment is not considered here. What is being considered is that the mechanical system has a structural (or *design*) entropy that makes the mechanical system work at less than 100% efficiency.

Giving strict consideration to non-living systems, work output depends on energy input. The system's ability to convert energy input into work output is a measure of the efficiency of the system. As a mechanical system operates, the output consists of the work and wasted energy, mostly in the form of heat. Much of this heat energy goes to increasing the entropy of the surrounding environment. Some of the wasted energy goes into overcoming the degrees of freedom in the system. The more degrees of freedom, the higher the system entropy, and the more energy required for a given amount of work.

Time and Energy

In production, time is a contributor to efficiency measures. Since productivity measures are time-dependent, one has to consider whether constricting available time will improve efficiency. If the timeframe is shorter, then typically more energy is required for the same unit of work, if entropy is not considered. Taking more time to produce more work will decrease productivity and increase to a greater extent the entropy of the production system. Also, as the entropy of the environment is continually increasing, then we know that operating a production system later may require more energy than operating it sooner. These differences are very small, but for very long time periods, or where time is dilated, the difference may be significant.

Energy and time are inversely related by the entropy of the system, so that if available increased energy needs, the amount of entropy determines what affect this will have on time, and if time increases, the amount of entropy determines how much of an energy change is possible.

If we think of a universe where the entropy is assumed to be constantly increasing, and we agree that in a work system entropy makes energy less useful, then ultimately the amount of energy required for a specific amount of work is inversely dependent on time. Applying energy at a later time will produce less work than applying it an earlier time to accomplish the same process.

On a universal scale, we could say that increased energy needs due to entropy are proportional to the increase of time, or that $E=kt$, where k is the constant of entropy expansion. Substituting this into Einstein's equation for energy, $E=mc^2$, gives us $kt= mc^2$ or $m = kt/c^2$ and $k = mc^2/t$. If time is measured in seconds and we substitute the speed of light as 300000meters/second then we surmise that k is inversely dependent on time raised to the third power.

~~~

Chapter VIII

Entropy and Society

Why should we consider the entropy of society? Is order important to the functioning, development and propagation of society?

We can define "society" as a system of individuals held together by a common norm. Whenever individuals, for whatever reasons, live in close proximity, they must become socialized, that is, accept rules that benefit and govern the interests of the society over their own individual interests, while protecting some of their own interests. Without this duality of purpose for societal rules, the individual will seek to flee the society or harm the society to make it collapse.

Most individuals are born into a society and most accept the rules of the system they are born into. Very few attempt to deviate from this acceptance, and those that do, either strive to change the rules, or are ostracized from the society, or are imprisoned by it. Societies depend on following rules in order to function and survive, and rules depend on order to be implemented and monitored. Rules (aka, laws) provide the behavioral norms for individuals in the society, and order provides the structure by which the rules are implemented and enforced. For these simple reasons, societies depend on a certain amount of order to flourish. Without the order that makes possible the exercise of rules, society devolves into bands of individuals with no cohesive norm.

Without some order, a society cannot function. With a lot of disorder, a society turns into a form of anarchy. Society makes up rules for the benefit of a plurality of its members. When individuals form a society, it means that they have set down some basic rules (that can become customs) on how they plan to coexist even though each member is ideally an independent free-thinking and free-willed person. By virtue of the need to coexist for emotional reinforcement and functional survival, humans design a system that has sufficient order to maintain and propagate rules, some of which are rules that self-servingly maintain and propagate order. Rules and order have formed a symbiotic relationship in society.

If we think of "society" as a system of individuals, then as in all systems, the society has entropy. The amount of entropy in this societal system is controlled by the rules and the adherence to them. Each individual is a subsystem of the societal system and each individual has its own entropy contribution to the society. The societal entropy is a function of the individual contributions and the algorithm (rules) that govern and define the society's functioning. Society as a system helps us understand from previous discussions that there are entropy thresholds below and above which the society will either not function or exist in its desired form. When the entropy thresholds are breached, the society either ceases to exist or it transforms itself into a different society with perhaps a different set of rules.

High-Entropy Societies

A society that is hypo-ordered becomes barely definable, almost lacking definition as a functioning collective system. If there are too few rules or if adequate rules are not enforced (i.e., non-functioning), the result is an anarchical state; and, an objective observer on the outside would discern few characteristics by which to define the society. The individual objects in the society display no definable behavior between themselves; they instead behave in apparent uncoordinated (or possibly random) fashion. This hypo-ordered society with very high entropy somewhat resembles a gas, where molecules each have their own properties and are definable, but the interaction between the molecules is random and mostly indefinable. The molecules' behavior is uncontrolled by the collective system, except that they are bound by it. A hypo-ordered society is a collective of individuals bound by geography or the basic need for survival but with minimal societal function. Generational progress in this type of society will be significantly more limited than in better ordered societies.

A society with high entropy is teetering on the verge of anarchy and is most likely in a state of confusion and uncertainty. Rules are present but the order necessary to enforce them is waning, and individuals may no longer know what is expected of them by the society. An example of such a high-entropy society is found in Russian history at the time of the Bolshevik revolutions and currently in many conflict zones. In a society with high entropy, the individual's role regains some preeminence over society, sometimes in order to survive, and other times in order to gain advantage. Motives aside, individual preeminence means more societal rules are not followed and a new plurality of individual rules further contributes to an increase in societal entropy. The higher the societal entropy, the more difficult it becomes to accomplish work within the societal framework, for example, the effort expended to get a license or to pick up and dispose of household waste. For these and other societal functions, extra energy and time is needed when there is an increase in societal entropy.

A society with high entropy also has a destabilizing psychological impact on the individual members. Order allows planning for expected outcomes which is essential to the conduct of enterprise. While very high societal entropy is not beneficial to societal work, the same may also be said for very low societal entropy. Society's benefits suffer at the extremes of societal entropy.

Low-Entropy Societies

Society benefits from a low entropy environment. Regulations and procedures are put in place to protect the populace and important functions are carried out on a timely basis with little concern that the functions will not be done. In a low entropy environment, funds are applied correctly, officials perform their duties and infrastructure is maintained. Maintaining a low societal entropy takes energy and thoughtfulness, but there is also a risk of reducing entropy too far below the desired equilibrium.

A hyper-ordered society is rigid and communication and energy travel along well-defined paths. Components of the society are not free to depart from their allocated roles. A lot is accomplished within the confines of the structure that is generally unyielding. Information is accumulated, stored and protected, but new information may not be generated in ideal quantities. Generational transfer of information is assured within the structure and there is little risk that this transfer will not occur.

However, at very low entropy, society can become less functional because its structure is too restrictive. Imagine, if you will, a society that has generated so many rules for the sake of more order that interaction is so regimented that it preempts accomplishment or innovation. Work is difficult to accomplish because the work function must follow so many rules (for example relating to federal regulation) and be executed in only a few specified ways.

This occurs when too many rules are made to "protect" societal goals, or to protect and preserve the group of individuals that make the rules. In physical terms, if we think of each rule as a steel girder in a high rise building, then we can imagine that too many girders (rules) will result in a decrease in occupancy space (function) because girders will be placed across habitable areas since they exceed the amount the structure actually needs to be viable. In this analogy, the building structure may be much stronger than it should be but the usefulness of the building has been diminished by the structure.

Another example to illustrate the point of too many societal rules would be putting the trash out for garbage collection. In a hyper-ordered society, only a certain shape of container with a certain color can be used, the container must be on the street and not on the property a foot away, and the handle must be pointing away from the curb. If all these rules are met, the trash will be picked up. If one of these rules (however well-intentioned their purpose) is not met, the trash is not picked up. Something as simple as collecting garbage can become burdensome to the point that the intended function suffers when too much order in the form of rules are put in place.

A further example of the disadvantages of hyper-ordered society is illustrated in the reasoning of Adam Smith. In Smith's time, society had built up a very rigid and controlled set of rules and laws to govern currencies and trade. Smith argued that these strict controls were counterproductive for the development of trade and wealth creation. Low entropy in a society *is* conducive to societal functioning, but very low entropy (hyper-order) creates rigidity and decreases functional output when deviations occur or are required. In Adam Smith's time, the entropy of the commerce subsystem of society had been reduced too far. A lack of degrees of freedom was stifling trade and the movement of goods. Entropy was reduced below the system's EE. Smith provided a recipe to revive trade and commerce by removing controls and relaxing rules, thus allowing more degrees of freedom and incentives for innovation. The benefit of Smith's arguments was to re-establish or shift the entropy of commerce closer to its desirable entropy equilibrium at which the system could be most productive.

Adam Smith also argued for the benefit to the greater prosperity of incentives from individual "profit". Profit is a necessary lubricant for work in a commercial system dependent on competition and one where entropy needs to increase to adjust functional production. Profiteering leads to greater entropy[31], which can be desirable if entropy is too low but undesirable if entropy is too high. Societal norms tend to restrict profit to what is considered reasonable. If profit is left unchecked, it moves too many resources into the hands of too few who nominally cannot effectively manage an abundance of resources. As a result, societal entropy increases because available energy is left unused (unproductive) and this energy goes to increasing entropy. Excess profit in most cases increases entropy while too little profit decreases entropy. As a lubricant for commerce, profit can be used to raise entropy to preferable levels.

Comparing Societal Entropy to Physical States

Societal variants can be paralleled with the three physical states of matter, mainly solid, liquid and gas. In this analogy, a hyper-ordered or

---

[31] An additional outflow of money brings additional degrees of freedom to the system, at a minimum.

very low-entropy society is comparable to the solid state, a well-ordered or entropy-efficient society to the liquid state, and a hypo-ordered or high-entropy society to the gaseous state.

In a hyper-ordered society, the objects of that society become like atoms in a crystal which can viabrate in place but have limited translational motion since they are bound by the structure of the crystalline lattice. To get an atom to migrate out of position requires a great deal of energy, so the expected tendency is no migration. When atoms are locked into a lattice framework, the crystal can be used as leverage, energy can pass through it, it can resonate or reflect, it can store vast amounts of knowledge, but inputs of energy produce a limited set of work outputs. Societies with very low entropy are similar to such a crystal.

The well-ordered society is analogous to a fluid state; there is some relationship and order between the components of the society, but not too much that it restricts the components from being flexible and translational. A lot is accomplished in this society, but the rate of accomplishment will vary with time due to the fluidity of the state, that is, as the structure varies over time, so does the efficiency of the society. Information generation is generally highest in this type of society because more components are interacting as they change positions and roles within the structure, but storage, protection, and generational transfer is generally less than in the hyper-ordered society because information transfer and storage pathways are not rigidly maintained.

The liquid or fluid state of matter is the most desirable analog for a well-functioning society. In the fluid state, relationships between objects in the system allow the system to have continuity and to perform desirable work without the rigidity of the solid state. Moreover, in the fluid state, objects can take new positions without disrupting the structure.

In the hypo-ordered society analogous to the gaseous state of water, the objects of the society have little or no relationships and making them do coordinated work is very difficult unless an outside force or boundary (container) is used to constrain the objects into performing definable work for the society. Societal objects may perform a great deal of "individual" work but this work does no translate necessarily to the benefit of society. The main societal contribution of the objects is "heat" or in sociological terms, "agitation". This is of marginal benefit and also may be harmful. Information generation occurs but it is not easily transferred or stored as a societal product and thus rarely propagated.

Societies do not usually start and end with the birth and death of a generation. The ability for society to outlive a generation allows for societal refinement in succeeding generations, and this comes from the society's ability to propagate knowledge about itself and about the rules that engender its nature. Knowledge propagation is crucial to the existence and lifespan of a society, and the accumulation and propagation of knowledge, whether written or oral, is heavily dependent in an inverse relationship to societal entropy. For transfer of information across generations, the solid state analogue is the most desirable because of the rigidity of processes. In the gaseous state analogue, society is generally in a state of anarchy which is not conducive to societal goals or to generation or transfer of information.

As societal entropy decreases, the efficacy of the society increases. The lower the societal entropy (up to a point), the greater is the progress of the society within a given time and energy frame. Lower societal entropy permits greater accumulation of knowledge and consequently more building on the constructs of previous generations, to benefit and propagate the society. However, like in all systems, each society has its own entropy equilibrium. The society functions best within the entropy band around this equilibrium. If you increase entropy too much, the society becomes anarchical and progress is slow. If you decrease entropy too much, the society begins to consume more resources than are necessary to make progress, and it becomes increasingly difficult to adopt new algorithms or to accommodate new needs. Governmental paralysis due to procedures and regulations is a good example of societal entropy becoming too low.

The importance of societal entropy and its connections to the propagation of knowledge are summarized in three hypotheses in Appendix C.

~~~

CHAPTER IX

ENTROPY AND INFORMATION

We use our minds to develop language so that the process of thinking can have an external physicality that allows thought to be transmitted asynchronously. Animals do communicate between themselves about their experiential knowledge but their languages do not appear to be structured enough to transmit their thought processes in codified form. The transcription of thought to the aural and visual senses is an entropy-increasing process and from the transient to codified symbols, it is an entropy reducing process. Language expressed in symbolic form allows developed ideas to be passed on from one generation to the next, bypassing the necessity for rudimentary biological encoding in our genes, aural or visual repetition, or exemplary repetition necessary to carry experiential knowledge forward. Successive refinement (or progress) is mostly possible through the communication of knowledge from generation to generation. The amount of work expended in one generation to conceive and codify knowledge does not have to be duplicated again and again in succeeding generations. The conception and development of the thought process has been blueprinted by putting it into words and symbols that another generation can understand, reuse and advance. The next generation will make use of the previous generation's progress without having to re-conceive it, and it may use this "knowledge" to develop a more sophisticated concept that is even more progressive and better at reducing entropy.

Knowledge Transfer

The transcription of thought to symbols to be carried across generations is the creation of knowledge. Knowledge creation is an entropy reduction process with progress as its byproduct. When the next generation is able to re-create the thoughts of the prior generation, then the order of that prior generation is carried immediately forward to the new generation. The sharing of thought from previous generations lowers entropy in the current generation.

As far as we know, animals apart from humans do not use symbols to transfer thought to succeeding generations, and so each new animal generation approaches the world with only its genetic instincts and experiential apprenticeship. In essence, each new generation must order their world anew by direct example or through instinctual genetic transfer. If the adult generation were to be lost such that their young are not raised by them, then this absence prevents experiential learning and the young animals have to rely solely on their instincts to proceed. Evidence of this is seen in adopted young animals, that when raised by humans, do not survive well as adults in the wild. Experiential learning in animals is an entropy reduction process that is repeated anew in each generation with undetectable forward progress.

One can conclude that progress in the animal kingdom is extremely slow in comparison to humans that benefit from codification of knowledge and societal continuity which provides for knowledge transfer. One might also conclude that this is the major factor that differentiates humans from the rest of the animal kingdom. When retained symbolic transfer of thought is not pursued, then each new generation faces their environment with essentially the same tools and knowledge as the previous generation. In such situations, the adaptation of most organisms to their surroundings results in painfully slow genetic and rote experiential transfer of that adaptation to new generations of organisms.

In human societies, each new generation inherits a knowledge base through the transfer of retained (recorded) thought (knowledge), which is apart from instinct or experiential learning. This knowledge transfer is the function of schools. To the extent that this function is performed well, a society can significantly reduce its entropy and make great progress.

The codification of societal law in all its complexities provides for generational stability in society. Social stability in turn provides the opportunities for refinement of existing knowledge and the opportunity and tools to discover and codify new knowledge. We have seen evidence of this in those societies, both eastern and western, that have codified knowledge.

In examining knowledge transfer, we should also include the means of codification (written symbols, music, art, etc.) and understand that this creates additional variations in knowledge transfer and entropy reduction. Knowledge transfer is not just limited to representative realism (representations of reality), but also to spirituality, feelings, and philosophy.

In primitive human societies where little or no codified knowledge exists, we do not see evidence of significant generational progress. In primitive human societies, an advance or discovery may be transmitted in experiential or vocal form to the younger generation, who at their own discretion and vicissitudes, may or may not transfer the new knowledge as adults to their young. Without codified knowledge transfer societal progress is very slow. The ability of humans to preserve thought in symbols and carry it forward as knowledge is a powerful reciprocal force that shapes our societal environments and their future sustainability.

The Information Tool

In previous sections we have discussed tools as a way to more efficiently reduce environmental entropy. Knowledge tools help with reductions in entropy. In the 20th century, a new human-developed[32] tool emerged to make possible the easier generation, manipulation and transfer of knowledge. This knowledge tool - the computer - is affecting the creation,

---

[32] As compared to nature-provided.

sharing, and transfer of knowledge as never before in recorded human history, and realizing incredible reductions in societal entropy.

The amount of human progress we will experience will be directly related to our preservation and transfer of knowledge. As a result, it is concluded here that because of computers, human society will experience progress that is many times faster and greater than in any previous recorded history. What implications does this now present for human societies? Can the rate of this progress outstrip society's abilities to manage and control this progress, or will this progress be self-limiting? Is knowledge codified by computers less permanent and more easily lost to future generations, therefore depriving them of generational progress? How is the individual and their ability to cope with change affected by abundant knowledge and rapid societal progression? These are just a few of the questions resulting from the widespread adoption of computers.

The computer tool has enhanced man's ability to *communicate* knowledge, so that the amount of knowledge transfer is far higher, faster and easier[33] than ever before. Because of this new "knowledge-ease", knowledge is no longer the sole property of a particular society or culture. Instead, it is rapidly becoming more the universal property of societies worldwide, transforming a heterogeneous set of societies and cultures into potentially one large global homogeneous society and culture. Homogeneity can bring about many positive and negative results to the human race as a whole.

Computer-based communication methods facilitate the transfer of knowledge so significantly by allowing persons in different cultures and areas of the world to exchange knowledge and also make personal one-to-one connections much more easily and quickly than ever possible before. Assisted by clever social software, this communication facility leverages opportunities for very large reductions in societal entropy while radically accelerating the amount of knowledge that is being produced, shared and retained. The computer as an information tool has set the stage for very large leaps of progress in technology, social interaction, and every human

---

[33] "easier" here means "with less barriers *and* importantly, less cost"

endeavor. Increased communication can lead to increased cooperation, reductions in fear, and meld cultures, erasing differences and blending the needs and wants of different societies. If we accept this thesis of cohesion from computer-enabled communication, we must acknowledge that we are poised for a new era in world politics, economics, and accelerating human progress.

The increasingly large gains in knowledge brought about by the "information tool" are also much more fragile than knowledge codified in previous generations and cultures. We still have some knowledge from ancient civilizations carved in stone, such that it has survived in codified form for thousands of years. Succeeding civilizations have provided us with knowledge codified on papyrus or on paper that has lasted two thousand or more years in good condition. Our current surge in knowledge creation and transfer is mainly codified in digital form on electronic memory systems that require the "information tool" to extract and interpret the stored knowledge. In all previous cultures before the twenty-first century, the tools used to create codified knowledge were not needed to later retrieve that knowledge. In our present century, we cannot look at a computer disc or "thumb-drive" and know what is encoded there-in without the aid of a computer (or perhaps by using sophisticated electron microscopes). The "information tool" is both the encoder and decoder.

Furthermore, many of our modern information storage systems are very fragile to their external environments. Unlike paper which can withstand sub-zero or boiling point temperatures, current digital media content can be lost due to high or very low temperatures, high moisture, magnetic fields, dust, or pulses in electrical power sources, just to name a few of many threats to content retention.

The computer tools we have built and rely on to codify and recover information are very complex and technologically sophisticated. Unlike the chisel and stone, ink and paper, or the printing press, the vast majority of people are unaware of the technical expertise and precision that are required to manufacture a microchip and the software that makes that electronic chip "smart". This expertise has become so decentralized because of its complexity, that no one company or organization in this century may be capable by themselves of recreating the tool that is needed to decode digital information.[34] The knowledge of the tool that we are using in our generation to transfer knowledge to future generations has become exceedingly specialized and divided across industries, and this specialization is compounded by the economic need to keep intellectual property (the methods, mechanisms, and algorithms) secret for competitive reasons. As a result, our digitized knowledge is far more vulnerable to being lost forever, that is, unavailable to succeeding generations, if certain conditions were to occur. The computer as an information tool is one of the greatest cultural entropy reducers yet one of the most fragile methods of generational knowledge transfer ever devised by man.

Digital Communication and Entropy

Claude Shannon in his first mathematical theory of communication proposed a five-part model, where one of the parts of communication is "noise". In later revisions to his theory, he equated "noise" to entropy, i.e., the amount of noise in the communication was a measure of the entropy of the message.

It is easy to understand that different forms of communication[35] have different levels of entropy or disorder associated with them. Digital communication has probably the lowest amount of entropy because the communication is quantized into discrete bits. By quantizing, damage to the communication is restricted to one or more bits, and the result

---

[34] In the 20th century, a few companies like IBM or NCR developed and manufactured by themselves the end-to-end systems and software for encoding and decoding electronic information. This is no longer the case in this century as manufacturing has moved to Asian countries while design and coding elements may reside in other countries. The modern-day computer, in whatever form it presents itself (personal computer, smartphone, tablet, systems controller, or supercomputer) is an amalgam of technologies form many companies and many parts of the world.

[35] Methods of communication, include hand signals versus auditory, handwritten versus the telegraph, amplitude modulation versus frequency modulation for radio, wired networks versus microwave, etc.

is a mechanism for communication with very little entropy. If none of the bits are changed or lost, then there is no noise associated with the communication. The entropy of the message that is communicated does not increase with the transmission distance as long as the discrete bits remain inviolate.

However, digitized communication does contain inherent entropy since it is often a discrete representation of a non-discrete phenomenon. Because most natural phenomena are not discrete, most digital representations of these continuous or analog phenomena have to be sampled, that is, measured at discrete intervals of time. The more numerous the intervals per unit of time, then the more exact is the digital representation, and the lower the inherent entropy. More measuring intervals per unit of time result in fewer parts of the analog phenomena being unmeasured, or indeterminate. The unmeasured parts can in theory take any value, though the likelihood (probability) is that they will have values similar to their measured intervals on either side. It is the uncertainty of unmeasured signals, though very small, that introduces one form of entropy into the digitized communication.

Discrete communication can be protected from incurring noise by the addition of error correcting algorithms. These are usually mathematical formula designed to relate all the bits in the message. These relationships are stored as additional bits in the message. The protection of the information content of a message relies not only on the algorithm but also on the extra relationship bits added to the message.

The error correcting algorithm is reducing the entropy of the communication system by creating a virtual structure around the message, but with the addition of more bits, it is also increasing the mass of the message. This is revealing since noise or increases in entropy of the communication are avoided by the application of a virtual relational structure and the addition of more "mass" to the communication. If one thinks of information bits as mass and content as weight, then one could derive classical and relative equations relating these values. To relate these values to time, we would add the speed of information transfer or how long it takes to send and receive the message.

What happens to a message when it has been compressed by an algorithm? In general, the more message information you have in disorder, the greater the entropy of the information, because with less order there are more unrelated elements, resulting in greater amounts of possible relationships between message elements. Since each possible relationship contains a discrete amount of entropy, overall entropy of the message is higher. Techniques similar to error correction are used to compress data in order to reduce the amount of bits (information mass) needed to communicate a message.

The result of compressing information is a decrease of information elements with more structure for the elements. As more structure is created, the amount of information elements can decrease the relationships are more tightly structured, that is, they are more strictly defined. The message has the same information content, but the compressed message has less entropy since there are fewer elements and more structure among the elements. In essence, physical entropy and information mass are reduced through virtual order imparted via the compression algorithm. A compressed message can be reconstructed in its entirety by reversing the compression process, removing virtual order, restoring information mass, and increasing message entropy.

This is not a primer on data compression, but rather a way to examine the property of entropy in the compression process. When information is compressed, the addition of message bit relationships creates structure in the message by which information can be condensed. The result is a message that is unintelligible without knowledge of the relationships or structure. As more structure is added, more entropy is removed from the message. The information is the same but the message has less entropy. Structure plays a key role in reducing system entropy.

If one continues to compress the information in a message, the relationships get more dependent on eachother, such that if a single element were to be lost, the entire message could be lost. In short, the tighter and more complex the structure (less degrees of freedom), the less is the entropy, but the greater is the rigidity of the system and the more susceptible the system becomes to unrecoverable damage.

In the physical world, such is the case with crystalline solids versus amorphous polymers. The polymers have more entropy but they can be dropped on the floor without loosing their properties or content, while shattering a crystal results in its original properties becoming mostly un-restorable. As structure increases, the result is less system entropy and increased system rigidity. There are likely ideal points of entropy where the benefits of less entropy are not outweighed by the fragility of rigidity. This also holds true for removing entropy from a message by compressing the content too far.

Can entropy then be held to a desirable constant by increasing a system's content such that this increase is structurally related to the system's content? Can we avoid entropy increases by applying structure and additional content? The answer from our digital information history is yes.

Virtual Entropy in Sender and Receiver

Message representations in communication can be very different things to sender and receiver. Verbally, the word "burro" means "donkey" to a Mexican, but an Italian standing next to him hears "butter". The fact that sounds alone do not carry the message, but that the intelligence or culture of the receiver determines the meaning of the message, must be considered. The framework or context of the sender and receiver is often as important as the message when communicating knowledge.

This idea is easily seen in the game of Bridge, where small phrases or words in the bidding process impart a large amount of information about the other players' cards, because of the intelligence of the sender and receiver (players) and the conventions (framework, relations, structure) in place at the time of the communication. The lower the entropy in the sender and receiver frameworks, the greater is the information that can be transmitted in a given amount of message bits. More structure in sender and receiver equates to more virtual order for the communication system.

With sufficient virtual order in sender and receiver, a single word can communicate as much information as desired. But a single word can also be lost or received in error. Though a message may contain a single word, the amount of information represented by that word may be very large, depending on the framework in which it is communicated. Likewise in such a framework, if a single word is lost, a large quantity of information is not delivered. This is somewhat paradoxical. If a large amount of information were represented by a message of fifty words, it is less probable that all these words would get lost and thus some of the information would be delivered in a "noisy" communication. Less entropy (more structure) in the message or in the sender and receiver provides greater accuracy and efficiency, but also greater rigidity and the risk of complete communication failure.

Thought and Communication

Thought has no mass in the traditional sense. It has no volume and it so it has no density. Yet to transmit thought conventionally, it must take a form we may generally refer to as "symbolic". Symbolic representations can be transmitted verbally or in writing. Auditory symbols can be spoken words, but they can also include grunts, squeals and other noises. Written symbols commonly use an alphabet or "phonetibet"[36] that are the building blocks of modern written language, or the picturegrams found in hieroglyphics or cuneiforms that represent one or more thoughts or objects. Thoughts therefore take a physical representation, be it written symbols, vocalized sound waves, etc.; and it is in these representations that entropy is introduced, first in the transcription where exact symbols may not exist to convey the exact thought, then in the process of transmission of the symbols, then in the process of reception, and finally in the process of interpretation (contextual translation), where the symbols become thought again. A certain amount of entropy is created each time a transition occurs

---

[36] My term for the symbols in Asian languages.

in the movement of thought. Thought conveyed with less entropy is more efficient than thought that is conveyed with more entropy in terms of the work performed in conveyance.

For the communication of thought to be very efficient, the right words must be chosen to transcribe the thought, the transmission must have very little noise, and the knowledge of the receiver must be such that the words re-create with fidelity the original thought. If these conditions exist, then thought has been transferred from one individual to another in a very efficient manner and with low entropy. Military communication during hostilities is a good example of efficient communication. Telegrams are another example. They both use unambiguous words and restrict the number of words in the message to minimize the effort and the error in reinterpretation back to thought.

Information and Knowledge

We have discussed knowledge and the communication of thought, and the entropy of communication. Now let us turn our attention more strictly to the subset of knowledge which we call "information". What is the difference between knowledge and information? Perhaps the best way to approach this subject is to first look at the etymology of the word "Information". Where does this word come from, how long has the concept of information been with us. What is information and what is knowledge? What is the difference between the two?

For the purposes of this analysis, information refers to a quantity of "knowledge-matter", from a single datum or bit to a larger quantity of data. Information is usually plural; it is a collection of datums. Information is the substance of "knowledge-matter", it has mass but is not necessarily knowledge. Information is required for knowledge, just as, for example, flour is the main ingredient for bread but it is not bread. A single symbol in the western alphabet can be a letter, but that one letter can also be a word, and also impart several meanings. A good example of this is the letter "i". Information can be a randomized collection of datums, or it can be a collection with some order applied to it. Order in information creates usefulness and meaning which engenders knowledge, and this is how this paper relates information to knowledge.

If we play an etymological game and separate "information" into "in formation", we are reminded that the way we think of the meaning of the word is related to order. Just as the soldiers of the 18th century were considered seasoned fighters only when they were "in formation", today we think of data as being useful only when it is "in formation", possessing some kind of meaningful order or structure. The higher the order of information, the greater is its potential as knowledge. Data without any order is random and is often referred to as "noise" because it has no value. Information, therefore, can be described as data with less entropy, and noise can be described as data with more entropy.

Knowledge is different from information in that it requires contextualization and the transfer of information. Data in some semblance of order becomes information. Information becomes knowledge when it is made relevant to a context or framework which impart further order in the form of relationships. Knowledge is transferred from one entity to another, and importantly, across generations. In this definition of knowledge, generational transfer must be possible and effectual. If knowledge is created and not transferred and stored, it reverts to information. Low information entropy is a key component of knowledge.

Information-Entropy Spectrum

What happens to data as its entropy gets very low or very high. If we assume that very low entropy means very structured data, we also assume that there is a limit at which data cannot be further structured, that is, all possible relationships between the datums have been incorporated into the information. Conversely, where data has so much entropy that its usefulness or knowledge-mass is negligible, there are few or no definable relationships between the datums. High-entropy data is very noisy and has very little value. As data becomes less ordered, the number of relationships between the datums become less and less until finally there are no relationships or structural content to connect the datums. When this happens, the data consists of individual datums, each with their own value, but there is no additional information in the data as a whole and minimal value. In physical terms, data at this point resembles an elemental gas, i.e., each datum is disassociated from others and their structural position is random.

We further conjecture that as data entropy reaches a very upper limit, the datums become further proximally disassociated. Not only are they random, but they begin to occupy a larger "data-space" that may become so large that at any given time, only a single datum is identifiable by observers within the space and time continuum, and all the other data are not within this observable range. In this case, the observable information that constitutes the reality of the observers is the value of a single datum, regardless of the quantity of data that originally occupied the observable data-space when its entropy was lower.

Without structure or relationships and with sufficient data-space, all the information in the data set is represented by a single observable value. At the entropy maximum, we could conjecture that even that one datum of information may disappear from an observer's space and time continuum. The datum would still be there, but the probability of observing it becomes so minimal, that in essence the datum has vanished. So at one extreme end of the information-entropy spectrum we have no data and no information.

At the other end of the spectrum where entropy becomes very low, the data become inter-connected to such a large degree that the information "knowledge-mass" increases significantly due a plethora of relationships. Structure helps to create mass. The greater the number of relationships between the datums, the greater is the value but also the more rigid is the structure holding the data. When all the relationships that are possible between the datums have been structured or added into the information, then the entropy of the data set is near a minimum. One can imagine the same knowledge in a data set having fewer datums if relationships and order increase between the data. In a minimum entropy data set with high virtual order, the knowledge can in theory be reduced to a very few datums.

If we put this discussion into a spectrum, we have the following:

Information-Entropy Spectrum

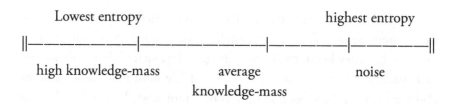

Usefulness of information depends on its relationships and structure, both measured by information entropy. We see that information is less usable at the endpoints of the entropy spectrum. Usable information requires some entropy, but not too much. We could classify "usable information" as a median entropy construct (MEC).

There is enhanced value in information with much lower entropy. Knowledge depends on lower information entropy. This is well known to computer scientists, particularly to database specialists, since they have been using related data within the structure of a database to increase the value of information. Knowledge provides for order restoration since without knowledge order may not be restorative.

Hyper-ordered information can be more completely understood and retrieved faster, and generate knowledge not previously obvious. A whole field of information science called "data mining" has sprung up from the ability to make data into more useful information by injecting additional order through classification structures and through making new relationships that were not part of the original information data set. This added order provides an increase in the information value of the data without changing the original content of the data. The added order can increase the data set size if it is embedded with the data (data bits + structure bits + relationship bits) which does affect efficiency. But the main point is the understanding that fully ordered information contains greater knowledge-mass. Like horses in gateposts at the start of a race, if the horses are not in the gateposts, they are of little value to the outcome of the race. By creating a tool to further order information, new relationships are possible between data in the information which we can consider virtual order since it is not part of the information data set. By establishing new relationships and querying those relationships, we learn new things from the same set of data. In short, this enhanced order creates more knowledge-mass and the information dataset becomes more useful.

Data Mining applications in recent years have yielded higher information value from existing datasets. Companies are mining sales and accounting information to discover relationships in the information that tell them more about their customers. Some companies have made a business of applying structure to numerous unrelated data sets to create information profiles on individuals. Each relationship that is established between elements in a dataset or between datasets lowers the information entropy. The lower the entropy (to a point), the higher is the value of the information.

Information Momentum

Usefulness of information depends on its relationships and structure, but it also depends on the speed at which the information is delivered. The speed at which information is delivered or processed will affect the information's entropy as well. This relativistic consideration means that the faster information is delivered, the lower the resultant entropy of the information. The slower information is delivered, the higher the resultant entropy. Just as structure can add information value to data, speed of information delivery can do the same.

A simple example illustrates this idea. Say that this week's edition of the *Economist* magazine which contains a lot of structured economic information is delivered electronically to you in one minute. The structure of the information is very high, the information speed is very fast, and consequently the information impact is very high. This information has a very great impact because it is ordered, useful, and timely enough to be applied to current economic opportunities you invest in. But if the same set of information in this week's *Economist* magazine were delivered to you very slowly, say one word per second, it would take one month at that rate to transfer about 20 text pages (a dataset of 43,200 words without pictures or diagrams) from the magazine. How applicable would this information be a month after it was published? The entropy of the *Economist* information dataset when received is the same as the entropy it had before it was transmitted (assuming no errors in transmission), but during the transfer, the virtual entropy of the data set to the observer or receiver is much higher if the speed of transfer is slow. As time elapses, frameworks change, which affect the virtual entropy of the information, and thus its resultant entropy. An increase in entropy renders the information less useful.

In short, Information Momentum determines the usefulness or impact of the dataset. We can correlate this to the equation for momentum which is equal to the mass of a physical object multiplied by its velocity and define Information Momentum (IM) as equal to the knowledge-mass of the dataset multiplied by its transfer velocity.

Structured Information and Freedom

The consequences of lowering information entropy to increase information value can be both good and bad for individuals and societies in general. Certainly more useful structured information advances societal progress, and less useful information delays progress. But more useful information can threaten society in several ways; the main threat being to individual freedom and liberty.

The greatest threat to personal freedom is the proliferation of structured information about a person and the person's surroundings. In prior human history, information about a person was available in many different forms, but the entropy associated with this information was very large in comparison to present day datasets.

Before written records, information entropy was so high that the dataset had considerable noise and lack of structure. That is why the verbal record of information required redundancy and repetition, so that effects of noise and minimal structure could be limited by correlating the verbal repetition. With the advent of the written record, pockets of

MEC information began to appear. The process of transferring the verbal record to written characters gave form and structure to information, reducing information entropy and eliminating some of the need for redundancy. Furthermore, the written record was structured enough to create knowledge-mass.

The advent of books allowed written records to be disseminated more widely and allow the reader to obtain value from relationships between books of information. The Bible with its many individual books is a good first example of this process. The Encyclopedias that came much later promised to further structure all known knowledge by categorizing and alphabetizing datasets of knowledge.

We now live in the golden age of information, or the "Information Age", so named because "mental machines" (computers) make it possible for more highly structured information to be more accessible in a shorter timeframe than ever possible before. A single Sunday newspaper carries in it more structured information than probably 100 years of all written records from the Bronze Age. The Information Age is possible because we have invented a machine particularly well-suited to process and deliver information. Just as the Industrial Age allowed physical work to be vastly amplified and accelerated by the use machines, the Information Age is amplifying and accelerating the mental processes of the human species. During World War II, it took a team of mathematicians solving differential equations many months to generate the structured data needed for firing projectiles from Navy battleships. Today, a single home computer can compute that same structured data in one hour or less and with no human assistance.

Whether it is calculating orbital trajectories or scanning your groceries, mental machines are processing huge amounts of information and vastly increasing as a result the amount of available information. Information momentum and knowledge mass have soared in the Information Age.

Information has been readily available in large amounts in the previous 200 years, but presently the ubiquity of the computer has made information more highly structured and its delivery very much faster, thereby increasing its impact. With the Internet, the mental machines of disparate localities are tied together in a world wide web, and this makes possible information that you want when you want it. Because of the relevancy provided by yet another set of mental machines ("search engines") and the immediacy of information provided by the Internet, the most trivial piece of knowledge can have a very high impact on individuals and societies due to its speed of delivery and its temporal relevancy.

Very structured information is very useful and very useful information about an individual can be very controlling. As you read this, your personal freedom is at the mercy of an electronic cloud that may soon become an electronic storm. The machines and communication channels are in place for personal information to be so structured and so quickly delivered that the existence of personal freedom (the ability to remain unencumbered by society) is virtually gone. Serious infractions of personal freedom have already occurred for many (e.g., identity theft disrupting an individual's life schedule and purchasing ability), but more serious situations exist on the horizon because the framework is already in place to allow information to be used as a tool for authoritarian control of individual lives and livelihood.

What started out as an information entropy reduction process has developed into a dangerous technological system that can force or coerce uniformity and control on human behavior, endeavor, and aspiration. The similarities to George Orwell's "1984"-like society are evident everywhere and while the framework for entropy reduction provides many benefits, it also contains within its structure the elements for abuse. There is nothing in the way except a moral reluctance or ethical imperative to prevent further abuse of personal freedoms through misuse of highly structured information. The infrastructure for enormous reductions in information entropy is already in place and this can be used for good as well as evil purposes.

Further increasing the impact of structured (low entropy) information on personal freedom is the speed at which information is transferred. More structured data increases knowledge-mass but communication speed additionally affects information momentum. With higher speeds of data transfer, information has higher impact. In the case of personal data, this impact can be of great benefit, such as in medical treatments, or of great threat, such as in denying an individual purchasing power. As information speed increases, information value increases, and information impact increases. A careful analysis of the impact of high-speed information on society would likely show that the loss of freedom and autonomy outweigh the benefits of the information.

So based on these simple observations, and the previous discussions, we can conclude with a number of hypotheses and theorems which are listed in Appendix B.

~~~

Chapter X
Entropy and the Mind

So what about the messy office desk? You've seen the one where there are piles of papers and folders in no apparent order in different stacks on top and around the desk. Is there order in this system? From an observational objective viewpoint we would say there is very little order. The desk system by itself does not allow an independent observer to step in and retrieve a particular paper or folder needed to accomplish work. The independent observer may perform a good deal of work to search the stacks of paper for the item that is needed, or the independent observer could perform a good deal of work to first decrease the physical entropy of the system in order to make finding a particular item less work. But unless some work is done by the independent observer, the messy desk has very little order and the work to retrieve a desired item is substantial and undesirable. To the independent observer, the desk system has too much entropy and the extra work performed will go to either using the system as it is or to reducing the system's entropy before using it. Let's look more closely at these two options.

Option 1: Using A System Without Reducing Its Physical Entropy

In the first option, the independent observer of a messy desk looks through stacks of papers for a desired document and will eventually find what is needed at an expense of an amount of energy that is based on the probability for finding the document in the search process within a given amount of time. In this search process, the messy desk system itself has not become more physically ordered, and there has not been a decrease in its entropy since the randomness of the papers was not changed during the search, except for the document that was recovered which now if placed on top of a stack is easily retrievable.

What could also happen when the observer searches for the document? While going through the stacks up to the point where the document is found, the observer could be registering mentally what has been looked at up to the point (predicted by probability) where the item is found. So if you were to ask the observer for a different item, the observer might be able to retrieve it immediately, if it was already seen during the search, and if its location in the stacks of paper is remembered knowledge in the observer's mind.

What has happened during the search for the document? First of all, the entropy of the system was not appreciably decreased by the search, but instead, a mental record of what was looked through during the first search has been registered in the mind of the observer. The observer's mind has therefore virtually decreased the entropy of the system by remembering the positions of some documents. So even though the physical system still has high entropy, new knowledge of the system allows a specific observer to create virtual order for the system resulting in an apparent entropy that is a result of the observer's interaction with the system and the observer's mental capabilities (memory & algorithms). If the observer is not independent, but is the owner of the desk or system and uses it frequently in its messy state, it is very likely that the observer's mind has already ordered the entire system virtually. This means that even though the desk has a high entropy to an independent observer, the apparent or virtual entropy for the system owner is much lower.

To an observer in another room that cannot see the physical order of the desk with its messy stacks of paper, but is limited to simply requesting and receiving documents, the apparent entropy of the system is based on the person that performs the retrieval task. If the person is independent and lacks knowledge of the system, the apparent entropy will be very high. But if the person is the system owner, the apparent entropy will be low because they can retrieve any paper quickly and hand it to the observer in the other room. In physical reality, the system's entropy is very high and the system is very inefficient to use. Including the virtual component of knowledge yields apparent entropy that is low. The mind has imparted virtual order on a disordered system and created a more ordered system.

As another example, assume that we have two desks and two desk owners with identical paper files on each desk. One desk has its paper files perfectly ordered and the other desk has stacks of paper files in no apparent physical order. If we were to put both desks and both desk owners behind a curtain so that we cannot see which is which, and then ask for a specific file, would we know which desk it came from? Our assumption might be that the file came from the desk with the most order. But we cannot know for sure. We presume the file came from the physically ordered desk but the apparent order of the disordered desk may have been higher.

We conclude that the mind can impart order to (and thus decrease the apparent entropy of) a physical system so as to make it more useful. This is both obvious and profound, because it extends our understanding of what intelligence can do with the physical world. It raises the consideration that different minds may order the same physical system differently and thus produce different virtual entropies for the same physical system. The virtual entropy depends on the observer's mental ability to induce order.

Different minds will create different levels of usefulness for the same physical system. But more importantly, when taken to its logical extension, we begin to understand that the perception of chaos is in fact just a perception, based on the observer's limited understanding and methods of order. Acknowledging this has ramifications in psychology, cognition, perception, mental health, and a host of other areas dealing with the interaction between human beings and their physical environment.

Option 2: Using A System By Reducing Its Physical Entropy

In the first option, discussed above, the search process using a messy desk system did not involve making the desk more physically ordered. In this second option for using the messy desk system, the person looking for a specific document performs work on the physical system to decrease the entropy of the system before starting the search for a desired item. Order is applied to the papers and folders by arranging them physically in such a manner that the initial and subsequent searches for documents are more quickly and efficiently completed. The physical entropy of the system itself has been decreased by advancing its structure such that the inanimate system becomes more efficient without the need for increasing its virtual order.

A certain amount of work goes into making a system more ordered and useful, and the result is that the system has lower entropy independent of the user, provided of course, that those who use it understand the physical order. This means, for instance, that if the documents on the desk are alphabetized, persons searching for a document would have to know the alphabet and the sequence of the alphabetic characters. There must be a pattern and an algorithm to follow for reductions in entropy to be useful.

By reducing the physical entropy of the system, all who subsequently use the system benefit from its decreased entropy. Of course, the system's entropy will increase with time and the system will become more disordered and less useful unless work is applied periodically to maintain the system's reduced physical entropy level.

Reductions in physical entropy are beneficial to an organization because they make the individual unnecessary for the system to be useful. Once one individual has worked to lower the entropy of the system, it remains low (though slowly increasing) even after the individual has departed. Achieving physical entropy reduction also frees the mind to concentrate on other things besides maintaining virtual entropy, and makes the individual available to work on other systems.

But physical entropy reduction is more restrictive than virtual entropy reduction. It takes a millisecond to mentally reposition something and virtual order is more flexible, allowing for changes to the structural pattern and ordering algorithm without the effort and time of physical rearrangement. This is easily demonstrated by considering the speed and flexibility of relational databases in delivering structured information.

In summation, imparting physical order to a system provides the benefits of reduced entropy to all those that later use the system. Low physical entropy however is not "free", work must be done to achieve it and maintain it. Reducing physical entropy requires time and energy.

Personal Entropy

When reductions in system entropy are virtual, the individual essentially becomes a part of the system in its reduced-entropy manifestation. Without the individual's presence and knowledge, the system can only be presented in its original higher entropy form. An individual then becomes indispensable to an organization as long as there are systems with virtual entropy in use by the organization. By the same token, individuals may be tied to systems and not given new opportunities in an organization because their value is linked to the resultant entropy of systems. At a more personal level, an individual who maintains virtual entropy may also feel burdened, or perhaps entitled, by this responsibility.

As a reminder of previous discussions, codification of thought is akin to physical entropy reduction in its benefits and constraints. We can define the process of putting "thoughts on paper" as a process of achieving physical entropy reduction. By codifying thought, virtual entropy is transferred from a mental form into a physical form that can transcend the individual. In taking physical form, the thought is now useful to many others. The author is no longer burdened by virtuosity, and may also experience the release of ownership and its responsibilities.

Perhaps this is why it is psychologically so difficult for some people to write, because when thoughts are codified in written form, they are freed from individual ownership and attain their own physical order, structure, and entropy. Once codified, the individual no longer has complete control over the thoughts and is no longer needed to sustain the apparent entropy of the uncodified thought.

The collective reaction of others to the codified thought also has a life of its own, separate and distinct from the author's control. When thoughts are retained by an individual, they remain under the control and service of only that individual, but when thoughts are recorded, they become the property and service of all who understand them.

The same is true for the artist. Until the thought is transferred to canvas or into musical notes, for example, it is both the interest and burden of the artist. Once presented physically, the artist who has made the transfer feels both release and apprehension, because the work is now freed from virtual order and attains physical structure which launches it into reality and opens it to admiration and criticism.

CONCLUSION

Entropy is the fabric of the universe. It connects all the forces and actions at play in the systems that make up nature and the physical and virtual realities of our existence. It is the last frontier of science, the understanding of which will answer many of the great questions of today and of the future and provide solutions to the problems that have eluded our understanding in the past.

This book is an effort to stimulate this understanding. It is hoped that the questions and hypotheses will generate new viewpoints on physical and mental processes and discussions of entropy in science, philosophy and religion that pave new avenues for investigation and discovery.

Einstein's search for a unified field theory was based on finding the principle(s) that connect the different forces of nature. This book is a launching point for an Entropy Framework of the universe with which we can more fully understand our reality and develop a unified theory of reality.

Appendix A: Hypotheses on Entropy

First: A system's nature is controlled and defined by its entropy equilibrium.

Second: All systems, both living and inanimate, tend toward their own individual entropy equilibrium, unless work is applied to maintain a different entropy equilibrium.

Third: A system's true nature occurs only at the system's entropy equilibrium point.

Fourth: If a system's entropy equilibrium changes enough to be outside its entropy band, the system's nature changes.

Fifth: The true nature of a system can be restored by reestablishing its original entropy equilibrium.

Sixth: The entropy of a closed system is directly related to the individual entropies of the subsystems that enable the closed system to do work.

Seventh: A closed system is able to do its designed work only within a bounded Entropy Band (EB), and is best suited for work at its Entropy Equilibrium (EE) within that band.

Appendix B: Hypotheses on Information Entropy

First: Information is not possible without some entropy.

Second: The amount of information in a data set is inversely
            proportional to the entropy of the data set within a
            moderate entropy construct.

Third: There is a limit to which data can be structured beyond
            which the information in the data set shrinks. This
            limit can be referred to as the "maxinf" (maximum
            information) limit. At this limit, a given data set
            achieves its maximum information value.

Fourth: The efficiency of the communication is inversely
            proportional to the amount of entropy present in the
            communication.

Fifth: To perfectly preserve information entropy for an
            observer in time-space, the speed of transfer must be
            instantaneous (unlikely).

Sixth: Information entropy increases during information
            transfer in time-space.

Seventh: As information transfer speed increases, observed
            information entropy decreases and information
            momentum (impact) increases.

Eight: If information transfer speed slows sufficiently, then
            information entropy will become high enough such
            that the information becomes noise to the observer.

Ninth: The speed of transfer at which information becomes noise is equivalent to the upper bound of the information entropy spectrum or e(I)=k/v(I).

APPENDIX C: HYPOTHESES ON SOCIETAL ENTROPY

First: A society is most effective when its entropy is at its equilibrium point.

Second: The creation and propagation of knowledge is inversely proportional to the creation and propagation of societal entropy.

Third: The propagation of knowledge is essential for the continuity of society.

Appendix D: Hypotheses on Virtual Entropy

First: A system has a real entropy but also an virtual entropy that depends on the observer's interaction with the system.

Second: A system's virtual entropy is a function of mental work (or intellect) applied to the system.

Third: The virtual entropy cannot become lower than the system's physical entropy threshold; otherwise, the system would be changed for the observer.

Fourth: The same system can have different virtual entropies.

Fifth: A system's virtual entropy is decreased by knowledge of the system and by the ability of the observer to virtualize the system, i.e., map it into the observer's intellect or imagination.

ENDNOTES

[1] The American Heritage® Science Dictionary Copyright © 2005 by Houghton Mifflin Company. Published by Houghton Mifflin Company. All rights reserved.

[2] http://en.wikipedia.org/wiki/Second_law_of_thermodynamics

[3] A distinction is made here between tidiness and orderliness. Tidiness refers to things being neatly positioned but it does not imply they are positioned in a particular order. In a messy room, things are not tidy but they can be arranged in a particular order that is not defined as "tidiness".

[4] Would a tidy and orderly room still feel creative? Probably it would feel less creative and more constructive or less friendly and more productive, or less warm and more sterile, less unique and more conformist. But these are human feelings attributed to inanimate objects and not necessarily scientific measures of a physical property. These feelings are not entropy but the result of entropy on the observer inside this closed system. In fact most "industrial" minds have been programmed so that the tidy clean room would remind them of work and the messy room would remind them of fun

[5] Other factors can include the energy available to the subsystems, the work algorithms, and the levels of freedom.

Printed in the United States
By Bookmasters